T0135507

On the determination and use of kinematic wavefield attributes for 3D seismic imaging

---

Bestimmung und Anwendung kinematischer Wellenfeldattribute bei 3D seismischen Abbildungsverfahren

Zur Erlangung des akademischen Grades eines

DOKTORS DER NATURWISSENSCHAFTEN

bei der Fakultät für Physik der

Universität Karlsruhe (TH)

genehmigte

DISSERTATION

von

Dipl.-Geophys. Steffen Bergler

aus

Isny im Allgäu

Tag der mündlichen Prüfung:	4. Juni 2004
Referent:	Prof. Dr. Peter Hubral
Korreferent:	Prof. Dr. Dirk Gajewski

Bibliografische Information Der Deutschen Bibliothek

Die Deutsche Bibliothek verzeichnet diese Publikation in der Deutschen
Nationalbibliografie; detaillierte bibliografische Daten sind im Internet über
http://dnb.ddb.de abrufbar.

ISBN 3-8325-0615-2

Logos Verlag Berlin
Comeniushof, Gubener Str. 47,
10243 Berlin
Tel.: +49 030 42 85 10 90
Fax: +49 030 42 85 10 92
INTERNET: http://www.logos-verlag.de

Abstract

In the last two decades a lot of effort has been directed towards methods that have the potential to succeed in imaging complex 3D subsurface structures from multi-coverage seismic data. The necessity of an estimate for the wave propagation velocities, required for transforming the data from the time domain to the depth domain, poses one of the fundamental problems in seismic imaging. Inadequate velocity models distort the final depth image. So-called data-oriented approaches are a class of imaging methods that avoid the explicit parameterisation of a velocity model in the first imaging steps. Instead, the data-oriented approaches parameterise the reflection events in the time domain and try to obtain as much information as possible from the measured data. The extracted information is then used to transform the seismic data into depth.

The common-reflection-surface (CRS) stack is one of the data-oriented imaging approaches. This method makes use of second-order traveltime approximations in order to describe seismic reflection events in the time domain. For the processing of data from a 3D acquisition, the traveltime equations can be used as stacking operators to simulate a zero-offset (ZO) volume of high accuracy and high signal-to-noise ratio from multi-coverage prestack data. During the stack, reflection energy from the entire five-dimensional data hyper-volume enters into the construction of one ZO sample. The eight parameters, which express the traveltime approximation for the ZO case, relate to kinematic wavefield attributes. These locally describe the propagation directions and curvatures of specific wavefronts at the Earth's surface which have travelled through the subsurface. Thus, the kinematic wavefield attributes constitute integral quantities of the medium's parameters and are suitable to estimate the properties of the Earth's interior. The accurate determination of the wavefield attributes is, therefore, a crucial step in the CRS processing.

In this thesis the derivation of the traveltime approximations is presented. The kinematic wavefield attributes are introduced by means of concepts known from geometrical optics. The determination of the eight kinematic wavefield attributes for the ZO case from 3D multi-coverage seismic data is elaborated. The applications of the attributes to support and facilitate 3D seismic imaging are discussed. In this context emphasis is put on the utilisation of the kinematic wavefield attributes for the 3D CRS stack. The proposed search algorithms are validated on a synthetic data example and have shown to be successful. Finally, the 3D CRS stack is applied to a real marine dataset. In this way the functionality of the search algorithms on complex data is verified. Moreover, the imaging quality of the 3D CRS stack is checked by migrating the simulated ZO volume to depth and comparing the obtained result with the result from a prestack depth migration. The comparison shows that the CRS based result is competitive to the result of the prestack depth migration. Thus,

CRS based imaging is an alternative to prestack depth migration due to the good imaging quality and also due to the provided information in form of the kinematic wavefield attributes.

Zusammenfassung

Vorbemerkung

Die vorliegende Arbeit ist bis auf diese Zusammenfassung in Englisch verfasst. Da auch in der deutschen Sprache einige englische Fachausdrücke gebräuchlich sind, wurde bei diesen Ausdrücken auf eine Übersetzung verzichtet. Sie werden, mit Ausnahme ihrer groß geschriebenen Abkürzungen, *kursiv* dargestellt.

Einführung

Reflexionsseismik

In der Geophysik wird zur Untersuchung geologischer Strukturen und/oder physikalischer Eigenschaften des Erdkörpers die Registrierung elastischer Energie in Form von seismischen Wellen genutzt. Dies ist möglich, da seismische Wellen sensitiv auf Änderungen der elastischen Eigenschaften des Erdinneren sind. Bei reflexionsseismischen Experimenten werden typischerweise künstliche Quellen, welche die seismischen Wellen anregen, an der Erdoberfläche eingesetzt. Empfänger, die in der Regel ebenfalls an der Erdoberfläche aufgestellt sind, messen dann die durch die ankommenden Wellen ausgelösten Bodenbewegungen (bei der Gewinnung von Daten an Land) oder Druckänderungen (bei der Gewinnung von Daten auf See) als Funktion der Zeit nach der Aktivierung der Quelle. Das Wellenfeld, das aufgrund einer Quellanregung mit einer Anordnung von Empfängern gemessen wurde, bildet ein *common-shot* Seismogramm. Dieses setzt sich aus einem Ensemble von seismischen Spuren zusammen. Eine Spur besteht bei einer digitalen Aufzeichnung aus einer diskreten Zeitreihe, die die aufgenommenen Signale darstellt. Ziel des reflexionsseismischen Experimentes ist es, vor allem diejenigen Wellen zu erfassen, die an den Diskontinuitäten im Untergrund reflektiert wurden. Je nach Art der Quelle können so Informationen über den Aufbau des Erduntergrundes von den ersten Metern bis zur tiefen Kruste und des oberen Erdmantels gewonnen werden. Reflexionsseismische Experimente werden vielfach von der Kohlenwasserstoffindustrie eingesetzt. Der Grund hierfür besteht darin, dass Erdöl- und Erdgaslagerstätten vornehmlich in Sedimentgesteinen bis zu 10 km Tiefe gefunden werden, zu deren Untersuchung die Reflexionsseismik besonders gut geeignet ist.

Bis vor 20 Jahren wurden in der Reflexionsseismik hauptsächlich 2D Experimente durchgeführt. Bei diesen werden Quellen an verschiedenen Positionen entlang einer Linie aktiviert und die angeregten Wellen jeweils von einer Reihe von Empfängern gemessen, die ebenfalls entlang dieser Linie positioniert sind. Durch Verschieben der Quellen und Empfänger entlang der Linie sollen Untergrundsstrukturen mit verschiedenen Quelle-Empfänger-Konfigurationen beleuchtet werden. Damit entsteht durch die so genannte Mehrfachüberdeckung eine Datenredundanz. 3D Experimente werden erst seit Mitte der 80er Jahre vermehrt von der Erdölindustrie eingesetzt. Dies liegt in der immensen Datenmenge begründet, die sich im Laufe eines 3D Experimentes anhäuft. Bei der Gewinnung von redundanten 3D Daten werden Quellen und Empfänger nicht nur entlang einer Linie, sondern auf einer gesamten Fläche verteilt und verschoben, was die Datenmenge gegenüber dem 2D Experiment vervielfacht. Bei einem Datenakquisitionsgebiet von $100 \, \text{km}^2$ sind heutzutage mehrere Terabytes an Daten, die gespeichert und verarbeitet werden müssen, nichts Ungewöhnliches. Die Bewältigung solcher Datenmengen ist erst mit dem starken Anstieg der Rechnerleistungen und Speicherkapazitäten der letzten Jahre möglich geworden.

Seismische Abbildungsverfahren zur Verarbeitung von 2D Datensätzen gehen fast alle stillschweigend von einer Zylindersymmetrie der Erde aus, wobei die Achse des Zylinders in der Horizontalen senkrecht zur Akquistionslinie verläuft. Abbilder des Erduntergrundes, die durch diese Verfahren erstellt werden, zeigen allerdings nur dann einen vertikalen Schnitt durch den wahren Untergrund, wenn diese 2D Annahme erfüllt ist. Bei komplexen dreidimensionalen Strukturen liefern 2D Abbildungsverfahren ein verzerrtes Bild des Untergrundes, das im schlimmsten Fall zu Fehlinterpretationen führt. In diesen Situationen können nur 3D Abbildungsverfahren die wahren Untergrundsstrukturen rekonstruieren. Da Erdöllagerstätten nicht selten in komplexen geologischen Gebieten vorkommen, in denen sich so genannte Fallen für Erdöl bilden können, wird in der Erdölindustrie zunehmend auf die 3D Datenakquisition und -verarbeitung gesetzt.

Datenorientierte Abbildungsverfahren und kinematische Wellenfeldattribute

Das Abbilden von komplizierten dreidimensionalen Strukturen aus seismischen Daten stellt eine große Herausforderung für jede Abbildungsmethode dar. Dabei hängt die Qualität des erstellten Tiefenabbildes immer maßgeblich von der Genauigkeit des benutzten Modells der Ausbreitungsgeschwindigkeit seismischer Wellen ab. Dieses Modell ist notwendig, um die Daten aus dem Zeitbereich (gegeben durch die Positionen von Quelle und Empfänger sowie der Aufnahmezeit nach Aktivierung der Quelle) in den Tiefenbereich zu transformieren. Verfahren, die dies bewerkstelligen, werden als Migrationsverfahren bezeichnet. Bei der so genannten *prestack* Tiefenmigration werden, indem man die Datenredundanz benutzt, das Erstellen eines Tiefenabbildes und die Konstruktion eines Geschwindigkeitsmodells mittels der Migrationsgeschwindigkeitsanalyse miteinander verwoben. Ausgehend von verschiedenen Akquisitionsgeometrien können durch die Datenredundanz gleich mehrere Abbilder der gleichen Untergrundsstrukturen mit Hilfe der *prestack* Tiefenmigration gemacht werden. Wurde dabei das richtige Geschwindigkeitsmodell benutzt, sind alle Abbilder identisch. Ist dies nicht der Fall, muss das Geschwindigkeitsmodell geändert und die gemessenen Daten erneut migriert werden. Dieser Vorgang wird so lange wiederholt, bis schließlich ein konsistentes Geschwindigkeitsmodell gefunden ist und die Tiefenabbilder somit (höchstwahrscheinlich) die wahren Strukturen im Untergrund zeigen.

Die Schwierigkeiten der 3D *prestack* Tiefenmigration liegen zum einen in der Bewältigung der riesigen Datenmengen (trotz der Möglichkeiten moderner Rechner) und zum anderen in der Parametrisierung eines geeigneten Startmodells für die Migrationsgeschwindigkeitsanalyse. Nur bei einem hinreichend genauen Startmodell ist tatsächlich gewährleistet, dass der iterative Prozess der Migrationsgeschwindigkeitsanalyse zum Ziel führt. Die Wahl eines Startmodells kann jedoch bei stark verrauschten Daten ein großes Problem darstellen.

Einen alternativen Weg bieten die so genannten datenorientierten Abbildungsverfahren. Diese Verfahren benötigen vorab keine explizite Parametrisierung der Ausbreitungsgeschwindigkeiten der Wellen in der Tiefe, sondern charakterisieren die gemessenen Reflexionsereignisse im Zeitbereich. Die hierzu notwendigen Parameter stellen integrale Größen dar, die die Information über denjenigen Teil des Untergrundes tragen, durch den die entsprechende Welle propagiert ist. Daher ist es möglich, aus den integralen Größen durch Inversion ein Modell der Ausbreitungsgeschwindigkeiten der Wellen abzuschätzen. Die in dieser Arbeit präsentierte *common-reflection-surface* (CRS) Stapelung gehört zur Klasse der datenorientierten Verfahren. Dabei werden Reflexionsereignisse, die zu verschiedenen Reflektoren in der Tiefe gehören, im Zeitbereich lokal durch eine Laufzeitapproximation zweiter Ordnung genähert. Zur Stapelung werden die Signale aufsummiert, die entlang der durch die Laufzeitapproximation beschriebene Fläche liegen. Das Stapelergebnis wird dann dem Punkt (*sample*) auf einer Spur zugeordnet, der dem Entwicklungspunkt der Laufzeitapproximation entspricht. Wählt man als Enwicklungspunkte *samples* von Spuren mit koinzidenten Quell- und Empfängerpositionen, so genannte *zero-offet* (ZO) Spuren, so ensteht ein ZO Volumen. Dieses kann zu einer ersten Interpretation der Strukturen im Untergrund herangezogen werden und reicht aus, um durch eine *poststack* Tiefenmigration ein Tiefenabbild zu erstellen. Der Rechenaufwand bei einer *poststack* Migration ist um ein Vielfaches kleiner als bei der *prestack* Migration, da durch die Stapelung die Menge der zu migrierenden Daten bereits stark reduziert wird. Die Parameter, die die Laufzeitapproximation beschreiben, können mit kinematischen Wellenfeldattributen in Verbindung gebracht werden. Diese charakterisieren die Propagationsrichtung und Krümmungen von Wellenfronten an der Erdoberfläche. Für das seismische Abbilden und die Dateninterpretation sind die kinematischen Wellenfeldattribute in vielerlei Hinsicht von Nutzen, unter anderem auch für die Erstellung eines Geschwindigkeitsmodells.

Für den 3D Fall werden in dieser Arbeit die theoretischen Grundlagen der kinematischen Wellenfeldattribute, deren Bestimmung aus seismischen Datensätzen sowie die Anwendungen der kinematischen Wellenfeldattribute bei Abbildungsverfahren, insbesondere bei der 3D CRS Stapelung, diskutiert. Diese drei Punkte sollen im Folgenden skizziert werden.

Theorie

Die Ausbreitung seismischer Raumwellen in elastischen Medien wird üblicherweise mit Hilfe der elastodynamischen Wellengleichung beschrieben. Diese besteht aus einem System dreier partieller Differentialgleichungen zweiter Ordnung. Für komplexe 3D Medien gibt es im Allgemeinen keine analytischen Lösungen für die Wellengleichung. Daher wurden verschiedene Methoden entwickelt, mit Hilfe derer man auch in komplexen Medien das seismische Wellenfeld untersuchen

kann. Eine dieser Methoden ist die Strahlenseismik, mit der es möglich ist, die Propagation aus-
gewählter Wellentypen in inhomogenen Medien zu beschreiben. Die Strahlenseismik basiert auf
einer Hochfrequenzapproximation der Lösung der Wellengleichung. In diesem Zusammenhang
bedeutet hochfrequent, dass die Wellenlänge einer betrachteten Welle klein gegenüber den cha-
rakteristischen Größen des Mediums sein muss.

Der Lösungsansatz für die Wellengleichung wird in der Strahlenseismik durch einen Reihenan-
satz, der so genannten *ray series*, beschrieben. Mit Hilfe dieser Reihe lässt sich unter anderem
die Eikonalgleichung herleiten, die die Laufzeiten von hochfrequenten Wellen bestimmt. Wendet
man die Methode der Charakteristiken auf die Eikonalgleichung an, so ergibt sich das (kinema-
tische) *ray tracing* System. Lösungen dieses Systems sind die Charakteristiken des Wellenfeldes,
die so genannten Strahlen. Die Laufzeiten der Wellen ergeben sich bei bekannter Verteilung der
Propagationsgeschwindigkeiten im Untergrund durch Integration entlang der Strahlwege.

Die Hamilton-Gleichung ist eine äquivalente Formulierung der Eikonalgleichung. Mit dem Kon-
zept der *surface-to-surface* Propagatormatrix (Bortfeld, 1989; Červený, 2001) werden im ersten
Abschnitt (Kapitel 2) des Theorieteils aus der Hamilton-Gleichung verschiedene Laufzeitnähe-
rungen zweiter Ordnung berechnet. Allgemein beschreibt eine Propagatormatrix in linearer Nä-
herung, wie sich die Dislokation und Richtungsabweichung eines so genannten Paraxialstrahls
im Verhältnis zu einem benachbarten Zentralstrahl aufgrund von Wellenpropagation ändert. Die
surface-to-surface Propagatormatrix betrachtet im Speziellen diese Verhältnisse an den Start- und
Endpunkten der Strahlen. Die Koeffizienten der hergeleiteten Laufzeitformeln sind durch die Ele-
mente der *surface-to-surface* Propagatormatrix beschrieben. Bei bekannter Laufzeit entlang des
Zentralstrahls lassen sich somit Näherungen zweiter Ordnung der Laufzeiten entlang von Paraxialstrahlen mit beliebigen Start- und Endpunkten, die auf einer ebenen Messoberfläche angenommen
werden, bestimmen. Im allgemeinen Fall sind dabei auch Start- und Endpunkte des Zentralstrahls
auf der Messoberfläche beliebig wählbar.

Im zweiten Teil der Arbeit (Kapitel 3) werden die Koeffizienten der Laufzeitformeln zu Attributen
von Wellenfronten in Beziehung gesetzt. Diese Wellenfronten werden durch verschiedene, zum
Teil hypothetische Experimente generiert. Besonders anschaulichen Charakter haben die Wellenfronten für den ZO Fall, das heißt, wenn die Start- und Endpunkte des Zentralstrahls zusammen-
fallen. Dann nämlich können die kinematischen Wellenfeldattribute von NIP (*normal-incidence
point*)-Welle und Normal-Welle benutzt werden, um die Laufzeitkoeffizenten erster und zweiter Ordnung zu beschreiben (siehe auch Höcht, 2002). In Abbildung 1 sind die Wellenfronten
der NIP- und Normal-Welle für ein einfaches Modell zu sehen. Die NIP-Welle wird durch eine
Punktquelle am Reflexionspunkt des ZO Zentralstrahls, der senkrecht zum Reflektor steht (daher *normal-incidence point*), angeregt. Sie propagiert entlang des Zentralstrahls und taucht an
dessen Start-/Endpunkt an der Oberfläche auf. Die Wellenfronten der NIP-Welle sind in Abbildung 1 zu verschiedenen Zeitpunkten durch die dunkelgrauen Flächen dargestellt. Die Normal-
Welle kann durch ein so genanntes *exploding reflector* Experiment angeregt werden. Bei diesem
hypothetischen Experiment wird ein Reflektorsegment in der Nähe des Reflexionspunktes des
Zentralstrahls mit Punktquellen bedeckt, die alle zum gleichen Zeitpunkt explodieren. Die daraus resultierende Welle ist die Normal-Welle, die ebenfalls entlang des Zentralstrahls zu dessen
Start-/Endpunkt an der Oberfläche propagiert. In Abbildung 1 sind die Wellenfronten der Normal-
Welle durch die hellgrauen Flächen dargestellt. Im theoretischen Teil dieser Arbeit wird gezeigt,

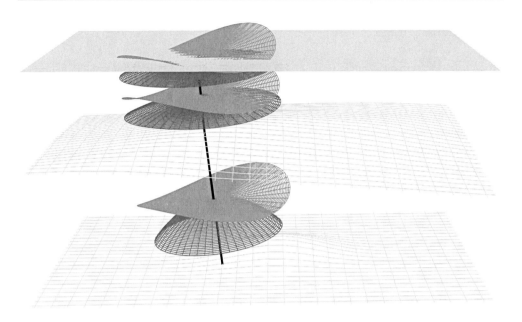

Abbildung 1: Modell mit zwei homogenen Schichten, die durch gekrümmte Grenzflächen getrennt sind. Die Wellenfronten der NIP-Welle (dunkelgrau) und der Normal-Welle (helleres grau), die entlang des Zentralstrahls (schwarz) propagieren, sind zu drei verschiedenen Zeitpunkten dargestellt.

dass die ZO Laufzeit des Zentralstrahls und die acht kinematischen Wellenfeldattribute, die die NIP- und Normal-Welle an der Oberfläche charakterisieren, ausreichen, um für den ZO Fall die Laufzeitapproximation für Paraxialstrahlen zu bestimmen. Die acht Wellenfeldattribute sind dabei durch folgende Größen gegeben:

- Die Propagationsrichtung der NIP- und Normal-Welle, die durch die Richtung des Zentralstrahls an der Oberfläche festgelegt ist. Die Richtung des Zentralstrahls ist dabei durch zwei Winkel bestimmt, die als Wellenfeldattribute in die Laufzeitformel eingehen.

- Die Krümmungen der NIP-Welle. Diese sind durch eine symmetrische 2×2 Matrix beschrieben. Somit liefert diese Matrix weitere drei Wellenfeldattribute zur Laufzeitformel.

- Die Krümmungen der Normal-Welle. Diese sind ebenfalls durch eine symmetrische 2×2 Matrix beschrieben und ergeben die letzten drei verbleibenden Wellenfeldattribute zur Laufzeitformel.

Implementierung

In den Implementierungskapiteln (Kapitel 4 und 5) geht es um die Bestimmung der kinematischen Wellenfeldattribute für den ZO Fall aus seismischen 3D Daten, vor allem im Hinblick auf die Anwendung der Wellenfeldattribute bei der 3D CRS Stapelung zur Erstellung eines ZO Volumens. Im 3D Fall ist der Datenraum durch die Laufzeit nach Aktivierung der Quelle und die Schuss- und Empfänger-Koordinaten der Spuren auf einer ebenen Messoberfläche gegeben und somit fünfdimensional. In diesem Datenraum beschreibt die Laufzeitapproximation eine Hyperfläche. Um die kinematischen Wellenfeldattribute für ein *sample* auf einer ZO Spur zu bestimmen, ist die Problemstellung nun, die Hyperfläche durch Variation der acht Parameter optimal an die Reflexionsereignisse in den Messdaten anzupassen. Dies wird durch Kohärenzanalysen entlang der verschiedenen getesteten Hyperflächen in den mehrfachüberdeckten Messdaten realisiert.

Die Suche nach den optimalen Wellenfeldattributen beinhaltet somit ein nichtlineares Optimierungsproblem mit acht Parametern. Eine simultane Optimierung der acht Parameter wäre aus theoretischer Sicht wünschenswert, ist aber aus Rechenzeitgründen vollkommen unpraktikabel. Daher werden in Kapitel 5 verschiedene Möglichkeiten vorgestellt, die die Rechenzeit gegenüber einer achtparametrigen Optimierung erheblich reduzieren, aber dennoch eine hinreichend genaue Bestimmung der Wellenfeldattribute gestatten. Die dazu notwendige Vorgehensweise kann an dieser Stelle nicht im Detail diskutiert werden, soll aber durch folgende drei Punkte zusammengefasst werden:

- Wie bereits von Müller (1999) und Mann (2002) für den 2D Fall vorgeschlagen, wird die gleichzeitige Optimierung aller Parameter durch mehrere aufeinander folgende Optimierungen mit weniger Parametern ersetzt. Dazu wird nicht der gesamte Datenraum gleichzeitig, sondern jeweils Teilmengen daraus benutzt.

- Die Anzahl der Spuren, die bei der Suche für ein ZO *sample* eingeht, wird so weit begrenzt, dass eine optimale Bestimmung der Attribute gewährleistet ist. Als Nebeneffekt wird durch diese Maßnahme auch die Suche beschleunigt.

- Die Wertebereiche der einzelnen Parameter werden so weit wie möglich eingeschränkt. Dazu werden zum einen Grenzen gesetzt, die sich aus physikalischen Überlegungen ergeben. Zum anderen können dabei auch Grenzen gesetzt werden, die sich aus Vorabuntersuchungen der Messdaten ergeben.

Da je nach Art der Datengewinnung (zum Beispiel Gewinnung von Daten an Land oder auf See) verschiedene Suchstrategien sinnvoll sein können, wird die Parameterbestimmung in Kapitel 5 auch unter dem Gesichtspunkt der verwendeten Akquisitionsgeometrie diskutiert.

Anwendungen

Die in Kapitel 5 vorgeschlagenen Suchstrategien wurden auf zwei Datensätze angewendet und die gefundenen kinematischen Wellenfeldattribute zur 3D CRS Stapelung benutzt. Kapitel 6 zeigt die

Resultate eines Tests an synthetischen Daten. Dieser Test wurde durchgeführt, um die Praktikabilität der implementierten Algorithmen zu überprüfen. Dabei haben die in dieser Arbeit entwickelten Suchalgorithmen sowohl in puncto Rechenzeit als auch in puncto Genauigkeit der erhaltenen Wellenfeldattribute ihre Anwendbarkeit unter Beweis gestellt. Der Vergleich zwischen dem ZO Volumen, das durch die 3D CRS Stapelung erstellt wurde, und dem ZO Volumen, das mit Hilfe des bekannten Untergrundmodells durch *ray tracing* zum Vergleich vorwärts modelliert wurde, zeigt, dass mittels der 3D CRS Stapelung korrekte ZO Reflexionsereignisse simuliert werden können.

Um die Durchführbarkeit der Suchalgorithmen und die Abbildungsqualität der 3D CRS Stapelung unter wesentlich schwierigeren Bedingungen als bei den synthetischen Daten zu überprüfen, wurden Tests an marinen Realdaten durchgeführt. Dieser Datensatz ist Eigentum der Firma *WesternGeco*, USA, die mir freundlicherweise erlaubt hat, meine Resultate für diese Daten im Rahmen meiner Doktorarbeit zu zeigen.

Die Ergebnisse der Tests (Kapitel 7) wurden in Zeit- und Tiefenbereichsresultate unterteilt. Erstere zeigen das ZO Volumen, das mit der 3D CRS Stapelung erhalten wurde, und die Volumina der zugehörigen kinematischen Wellenfeldattribute. Das ZO Volumen offenbart bereits deutlich einige Eigenschaften des zu untersuchenden Gebiets. Zum einen sind Reflexionsereignisse zu erkennen, die eindeutig von weitgehend söhligen Sedimentstrukturen stammen. Zum anderen beobachtet man ein starkes Reflexionsereignis, das auf einen großen Sprung der elastischen Parameter zwischen der Schicht oberhalb des damit assoziierten Reflektors und der darunterliegenden Schicht schließen lässt. Dieser Reflektor ist, wie man später an den Tiefenabbildern erkennt, die Oberfläche eines Salzkörpers. Das starke Reflexionsereignis wird von vielen Diffraktionsereignissen überlagert, die auf Rauigkeiten der Salzoberfläche hinweisen. Diese sind relativ gut durch die großen Werte in dem Attributvolumen, das den Auftauchwinkel des Zentralstrahls zeigt, erkennbar. Darüber hinaus lassen sich mit den Wellenfeldattributen multiple Reflexionsereignisse identifizieren, die zu Wellen gehören, die mehrfach reflektiert wurden. Die meisten seismischen Abbildungsverfahren arbeiten nur mit Primärreflexionen, die zu Wellen gehören, die nur einmal reflektiert wurden. Daher gilt es, die multiplen Reflexionsereignisse frühzeitig während der Datenverabeitung zu ermitteln, um Fehlinterpretationen zu vermeiden. Einige multiple Reflexionsereignisse verraten sich durch ihre im Vergleich zu benachbarten Primärreflexionen niedrigen Stapelgeschwindigkeiten. Letztere sind Größen, die sich aus den Wellenfeldattributen ableiten lassen (siehe Kapitel 5). Neben der ersten Interpretation des Untergrundes wurde das durch die 3D CRS Stapelung erstellte ZO Volumen mit einem von der 3D *common-midpoint* (CMP) Stapelung produzierten ZO Volumen verglichen. Die CMP Stapelung ist ein innerhalb der seismischen Datenverabeitung etabliertes Verfahren, bei dem zur Konstruktion einer ZO Spur wesentlich weniger Spuren der mehrfachüberdeckten Messdaten eingehen als bei der CRS Stapelung. Die Tatsache, dass in die CRS Stapelung mehr Spuren eingehen, spiegelt sich deutlich in der Unterdrückung von ungewolltem Rauschen wider. Das führt letztlich zu einer besseren Identifizierbarkeit und Kontinuität der Reflexionsereignisse durch die CRS Stapelung.

Das Ergebnis der 3D CRS Stapelung wurde für eine *poststack* Tiefenmigration verwendet. Zur Evaluierung der Abbildungsqualität dieser *poststack* Tiefenmigration wurde zusätzlich eine *prestack* Tiefenmigration durchgeführt. Damit der Vergleich möglichst objektiv ist, wurde für beide Verfahren das gleiche Geschwindigkeitsmodell benutzt. Des Weiteren wurde darauf geachtet, dass die gleiche Anzahl von Spuren aus den Messdaten für das Erstellen der Endresultate herangezogen wurde. Vertikale Schnitte der Tiefenabbilder, die mit beiden Verfahren erhalten wurden, sind in Abbildung 2 zu sehen. Diese zeigen klar die Oberkante und Teile der Unterkante eines Salzkörpers. Vorteile der *poststack* Tiefenmigration sind zum einen in der größeren vertikalen Auflösung zu sehen, die sich vor allem durch die bessere Identifizierbarkeit der dünnen Schichtung im oberen Teil der Bilder äußert. Zum anderen zeigen sich an den Flanken des Salzkörpers Reflektorabbilder oder Teile von Reflektorabbildern, die in der *prestack* Tiefenmigration nicht ohne Weiteres zu erkennen sind. Ein etwas besseres Abbild eines Systems von Verwerfungen zeigt sich bei der *prestack* Tiefenmigration im oberen rechten Teil der Bilder. Dies lässt sich auf die Komplexität der Wellenfronten durch die Verwerfungen zurückführen. Die zugehörigen Reflexionsereignisse können nicht mehr ohne Weiteres durch Funktionen zweiter Ordnung beschrieben werden, was zu Einbußen in der Abbildungsqualität bei der *poststack* Migration führt. Zusammenfassend kann jedoch gesagt werden, dass die *poststack* Tiefenmigration ein zur wesentlich aufwändigeren *prestack* Tiefenmigration konkurrenzfähiges Resultat ergeben hat, das in manchen Teilen sogar besser ist. Dies wiederum zeigt, dass die Anwendung der 3D CRS Stapelung erfolgreich war und die kinematischen Wellenfeldattribute stabil bestimmt wurden.

Schlussfolgerungen

Die kinematischen Wellenfeldattribute, mit deren Hilfe Laufzeitformeln zweiter Ordnung beschreibbar sind, können ohne explizite Kenntnis des Geschwindigkeitsmodells aus reflexionsseismischen Datensätzen bestimmt werden. Die in der Arbeit vorgestellten Suchalgorithmen zur Bestimmung der Wellenfeldattribute haben sich sowohl bei synthetischen als auch bei realen Daten als praktikabel erwiesen. Der Einsatz der Laufzeitformeln als so genannte Stapeloperatoren für die 3D CRS Stapelung zeigt, dass diese zur Erstellung von ZO Volumen sehr gut geeignet sind. Die Qualität der Tiefenabbilder, die durch die *poststack* Tiefenmigration erstellt wurden, ist durchaus mit der Qualität der Tiefenabbilder einer *prestack* Tiefenmigration vergleichbar, in manchen Teilen sogar besser. Es ist zu erwarten, dass sich die Qualität der CRS-basierten Abbilder noch deutlicher von der *prestack* Tiefenmigration abhebt, wenn die Daten stark verrauscht sind. In diesem Fall ist es besonders schwierig, mit Hilfe der Migrationsgeschwindigkeitsanalyse ein Geschwindigkeitsmodell zu bestimmen. Die 3D CRS Methode hat dagegen aufgrund der starken Rauschunterdrückung durch die Stapelung das Potenzial, bei verrauschten Daten immer noch gute Tiefenabbilder zu liefern. Grund dafür sind auch die kinematischen Wellenfeldattribute, mit denen ein Geschwindigkeitsmodell zur Tiefenmigration abgeschätzt werden kann. Liefert die Stapelung ein gutes Ergebnis, kann davon ausgegangen werden, dass die Wellenfeldattribute stabil bestimmt wurden und daher ein korrektes Geschwindigkeitsmodell aus den Wellenfeldattributen abzuleiten ist. Somit kann man schlussfolgern, dass die 3D CRS Methode und die Wellenfeldattribute das seismische Abbilden in bestimmten Fällen deutlich vereinfachen können und eine gute Alternative zu vielen anderen seismischen Abbildungsverfahren darstellen.

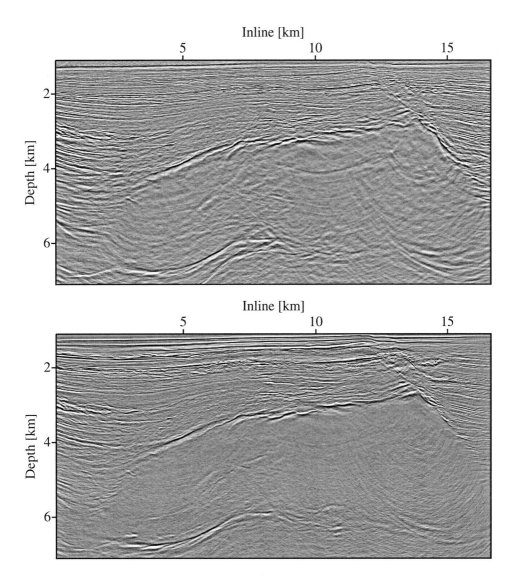

Abbildung 2: Vergleich zwischen der *poststack* Tiefenmigration (oben), die nach der 3D CRS Stapelung durchgeführt wurde, und der *prestack* Tiefenmigration (unten).

Contents

Chapter 1

Introduction

This thesis describes the determination of kinematic wavefield attributes from 3D seismic data in the context of the 3D common-reflection-surface (CRS) stack. The present chapter is intended to introduce the reader to the subject of 3D seismic imaging, thus motivating the interest in kinematic wavefield attributes for imaging purposes.

1.1 The seismic reflection experiment

A popular way to gather information about the Earth's interior, ranging from a few metres for engineering purposes down to the deep crust and the upper mantle, is to investigate the propagation of elastic energy in the form of seismic waves from seismic reflection experiments. For this purpose controlled sources and receivers are usually spread out on the Earth's surface. Due to heterogeneities of the elastic properties in the subsurface, wave phenomena such as reflection, diffraction, and refraction, influence the seismic waves during propagation. Thus, information about the subsurface properties can be deduced from the backscattered wavefield detected at the receivers. In reflection seismics mainly so-called primary reflections (recorded signals corresponding to waves which were reflected only once) contained in the wavefield are utilised for this objective.

Seismic reflection experiments are carried out on land, at sea, or the transition zones in between. In land data acquisition explosives, vibrators, or weight drops are, for instance, used as seismic sources. In this case the receivers (geophones) measure one or more components of the particle motion (or some quantity from which the particle motion can be derived) due to the excitation of seismic waves as a function of the time after activation of the source (traveltime). In marine data acquisition airguns are frequently employed as sources which release highly compressed air into the water. Marine receivers (hydrophones) then measure the variation of the pressure as a result of the emerging seismic waves as a function of the traveltime. During the 1990s, the marine experiment was extended by deploying receivers on the sea floor, enabling to record a multi-component wavefield in a marine setting. Usually, seismic sources are activated at a multitude of locations on the Earth's surface, where for each shot an array of receivers is used. The recorded wavefield

1

at the receivers due to the activation of one source makes up a common-shot seismogram. The ensemble of many common-shot seismograms, in turn, represents a multi-coverage dataset.

Seismic reflection experiments are the principal way oil and gas companies find new reserves and exploit and develop existing resources. The reason for this is that hydrocarbon reservoirs are typically found in sedimentary rocks with layered structures which reflect the seismic waves very well. These stratigraphic reflectors are often found to be slightly curved interfaces between different sedimentary layers or units. However, sedimentary rocks can also exhibit complex structures due to later tectonic events, such as folding, faulting, and intrusion of salt or basalt. Such geologic settings are in many cases the most interesting part of the subsurface because these may form stratigraphic traps which serve as a sealed geologic container capable of retaining hydrocarbons. The recorded wavefield associated with complex subsurface structures is necessarily also complex. Therefore, the processing and interpretation of such data are difficult tasks and remain a challenge.

1.2   3D seismic imaging

Unprocessed seismic data often provide a coherent picture of structures. However, this picture does not give a proper subsurface image but suffers from several distorting effects. Seismic imaging aims at correcting these effects, thus creating a representation of the subsurface properties. In fact, seismic imaging includes a culmination of processes, such as deconvolution, static corrections, velocity analysis and stacking, leading up to migration (Yilmaz, 2001a). The latter process maps the recorded wavefield into the subsurface to produce a proper image of the reflectors and is performed primarily for structural imaging purposes only (Gray, 2001). Nowadays, the amplitude information of migrated images is also increasingly used to predict rock properties.

For a long time only 2D processing algorithms were applied to the seismic data which were acquired using sources and receivers deployed along a more or less straight line on the Earth's surface. Many 2D seismic imaging schemes inherently assume that the Earth is a cylinder with its axis pointing perpendicular to the acquisition line. If this assumption is fulfilled, the 2D seismic image obtained after migration can represent a true section of the subsurface. However, if this assumption happens to be invalid, 2D seismic imaging may produce a distorted image of the subsurface which leads to misinterpretations. Moreover, it has become increasingly apparent that subsurface models resulting from mapping with 2D seismic data lack the detail obtained from the 3D methods. In case the complex geological structures are of a three-dimensional nature, only 3D seismic imaging methods are able to provide an accurate subsurface image leading to a more reliable interpretation. The proliferation of computing power during the last 15 years makes many 3D seismic imaging methods meanwhile feasible. Therefore, 3D seismic data acquisition as well as processing are predominately carried out in the marine case and, where possible, replace more and more 2D acquisition and processing for the land case. Data examples emphasising the improvements of the final subsurface image due to the use of 3D seismic imaging methods are shown in Yilmaz (2001b) and Biondi (2003).

1.3 Data-oriented imaging and kinematic wavefield attributes

For any migration process mapping the reflection events detected at traveltime t to particular sub-surface locations, there is the need for a velocity model which describes the propagation velocities of the seismic waves. Therefore, the quality of the depth migration strongly depends on the accuracy of the used velocity model. Unfortunately, apart from the near surface and along boreholes, the true distribution of the velocities inside the Earth is not known beforehand. Thus, it is necessary to rely on estimates for the velocities inside the Earth. The parameterisation of these estimates can become an extremely difficult problem. This applies especially in the case of complex data, where inadequate acquisition leads to irregular data geometries and poor data quality. For this reason imaging methods termed data-driven or data-oriented seismic imaging methods have been devised which rather try to parameterise the traveltimes of recorded reflection events instead of the velocity model in depth. Moreover, these methods attempt to do most of the processing in the time domain and extract as much information as possible from the seismic data before migration.

From my point of view, the terms *data-driven* and *data-oriented* for the above described way of processing may be misleading. Approaches like prestack depth migration, which directly migrate the acquired data to depth, are often called model-based methods. In fact, these methods are also driven by the data. However, the model-based approaches rely on the accuracy of some initial estimate of the velocity model which needs to be known beforehand and which has an influence on the final result.

There are quite a few methods which follow the data-driven approach. For example, the methods which utilise multi-parameter traveltime approximations of the reflection events as, for instance, the Polystack (de Bazelaire, 1988; Thore et al., 1994; de Bazelaire and Viallix, 1994), the homeomorphic imaging method (Gelchinsky and Keydar, 1999) which led to Multifocusing (Gelchinsky et al., 1999; Landa et al., 1999), and the common-reflection-point stack (Höcht et al., 1997; Perroud et al., 1997). Another approach which falls into the data-driven category is the Common Focal Point (CFP) method (Berkhout, 1997a,b; Bolte, 2003). This method extracts from seismic data so-called CFP operators which describe the one-way traveltimes from a subsurface grid point to the Earth's surface. The CFP operators are useful for various imaging purposes, for instance, the velocity model estimation (Cox et al., 2001).

The CRS stack (see, e. g. Müller, 1999; Mann, 2002, for the 2D case) described in this thesis is a data-driven imaging method which also makes use of a multi-parameter traveltime formula. This formula approximately describes reflection events in the full multi-coverage data (hyper-)volume. Initially, it was solely applied to 2D datasets (Mann et al., 1999; Jäger et al., 2001) to construct a zero-offset (ZO) section from multi-coverage data, where the ZO section corresponds to a measurement with coinciding source and receiver pairs. For simpler subsurface media, the ZO section is already suitable for a first interpretation of the subsurface structures. As a result of the CRS stack, the signal-to-noise ratio of the ZO section is increased while the amount of data is decreased compared to the measured multi-coverage prestack data. Due to the increase of computing power in the last years, there are meanwhile first applications of the CRS stack to 3D datasets as well (Cristini et al., 2002; Bergler et al., 2002). In this case entire ZO volumes are produced from the multi-coverage data.

Firstly, the benefits of the CRS stack were mainly attributed to the improved imaging quality in case of a low signal-to-noise ratio and complex subsurface structures compared to other methods which construct ZO sections (Mann et al., 1999; Müller, 1999). In the course of this thesis the reasons for these improvements are discussed. Meanwhile, 2D data examples were shown, where CRS based processing also led to depth images which were competitive or even better than results from prestack depth migration (Trappe et al., 2001). Such comparisons, yet of 3D data examples, will be shown later in this work. In the last years more and more the kinematic wavefield attributes extracted from the seismic data during the CRS processing come to the fore. These kinematic wavefield attributes describe the emergence direction and wavefront curvatures of specific waves at the Earth's surface which propagated through the subsurface. Due to the sensitivity of the kinematic wavefield attributes to changes in velocity, these attributes are integral properties of the true velocity distribution in depth and, therefore, suitable for the construction of an estimate for the velocity model. This application, as well as further applications for seismic imaging, make the accurate determination of the kinematic wavefield attributes a crucial step during the CRS processing. Therefore, the determination of the attributes from seismic data is elaborated in detail in this thesis.

1.4 Outline of the thesis

The structure of the thesis reflects the natural progression of the work. Chapter 2 starts with the description of the kinematic aspects of high-frequency wave propagation which involves the calculation of traveltimes along rays. The aim of the chapter is to formulate second-order traveltime approximations valid along rays with arbitrary starting and end points in the vicinity of a central ray by means of the surface-to-surface propagator formalism (Bortfeld, 1989; Schleicher et al., 1993; Červený, 2001). In Chapter 3 the kinematic wavefield attributes are introduced and related to the coefficients of the second-order traveltime approximations derived in Chapter 2. This is accomplished using concepts of geometrical optics (Höcht et al., 1999; Höcht, 2002). Chapter 4 summarises the commonly employed 3D acquisition schemes and discusses different seismic imaging processes. It is outlined how the 3D CRS stack and the kinematic wavefield attributes can support the imaging methods applied nowadays or even be an alternative to these. The determination of the kinematic wavefield attributes from 3D data is explained in Chapter 5. For this purpose, the search algorithms devised in this thesis will be introduced which take into account both, the accuracy of the attributes and the required computing time. In the 3D case the latter is an important issue to be considered in order to make the whole 3D CRS processing feasible. In Chapter 6 the applicability of the search algorithms proposed in Chapter 5 are validated for a synthetic data example. The performance of the 3D CRS stack and the search algorithms on real data from 3D marine acquisition is tested in Chapter 7. Moreover, the image quality of the 3D CRS stack results is evaluated by migrating these to depth and comparing the obtained images with results from a 3D prestack depth migration. The Appendices A - D refer to the theory chapters (Chapters 2 and 3). Appendix E discusses a 3D time migration approach based on the kinematic wavefield attributes.

1.5 Notation

In order to simplify the reading of the thesis, i. e. mainly of the formulas, I shortly summarise in this section the notation and the coordinate systems used throughout the thesis.

1.5.1 Scalars, vectors, and matrices

All vectors and matrices are bold in order to distinguish these quantities from scalar variables; vectors are written in small letters and matrices with capital letters. 4×4 matrices are underlined, 3×3 matrices and 3D vectors are marked with a hat ($\hat{}$) to differentiate these quantities from 2×2 matrices and 2D vectors. Examples for the vector and matrix notation are given in Table 1.1. Different operations applied to vectors and matrices are listed in Table 1.2.

Vectors			Matrices		
	Symbol	Element		Symbol	Element
2D	$\mathbf{a}$	a_i, (i = x,y)	2×2	$\mathbf{A}$	a_{ij}, (i,j = 0,1)
3D	$\hat{\mathbf{a}}$	a_i, (i = x,y,z)	3×3	$\hat{\mathbf{A}}$	a_{ij}, (i,j = 0,1,2)
			4×4	$\underline{\mathbf{A}}$	a_{ij}, (i,j = 0,1,2,3)

Table 1.1: Notation of vectors and matrices. The notation of the elements of vectors differs in some cases which are then explicitly mentioned.

Vector operations		Matrix operations	
Operation	Meaning	Operation	Meaning
$\cdot$	inner product	$\mathbf{A}^{-1}$	inverse of $\mathbf{A}$
$\mathbf{a}^T$	transpose of $\mathbf{a}$	$\mathbf{A}^T$	transpose of $\mathbf{A}$
		$\mathbf{A}^{-T}$	inverse of $\mathbf{A}^T$

Table 1.2: Vector and matrix operations.

Moreover, the notation $\frac{dt}{d\mathbf{a}}$ is often used which stands for

$$\frac{dt}{d\mathbf{a}} = \begin{pmatrix} \frac{\partial t}{\partial a_x} \\ \frac{\partial t}{\partial a_y} \end{pmatrix}. \tag{1.1}$$

1.5.2 Coordinate systems

In the course of the thesis three different Cartesian coordinate systems are used. These systems, as well as variables referring to these systems, are listed in Table 1.3.

System	Axes	Scalar, vector, matrix referring to the system
Global system	$(\hat{\mathbf{e}}_x, \hat{\mathbf{e}}_y, \hat{\mathbf{e}}_z)$	$a, \mathbf{a}, \hat{\mathbf{a}}, \mathbf{A}$
Local system	$(\hat{\mathbf{e}}'_x, \hat{\mathbf{e}}'_y, \hat{\mathbf{e}}'_z)$	$a', \mathbf{a}', \hat{\mathbf{a}}', \mathbf{A}'$
Local ray-centred system	$(\tilde{\hat{\mathbf{e}}}_x, \tilde{\hat{\mathbf{e}}}_y, \tilde{\hat{\mathbf{e}}}_z)$	$\tilde{a}, \tilde{\mathbf{a}}, \tilde{\hat{\mathbf{a}}}, \tilde{\mathbf{A}}$

Table 1.3: Coordinate systems used throughout this thesis.

Chapter 2

From the eikonal equation to second-order traveltime approximations

In reflection seismics, apart from the direct vicinity of the seismic source, the Earth is assumed to behave like an elastic continuum for small deformations. Therefore, classical continuum mechanics with a linear stress-strain relation provides the mathematical background to establish equations which govern the seismic wavefield. This system of equations consists of three partial differential equations of the second order and represents the elastodynamic wave equation. For complex 3D media analytical solutions to the elastodynamic wave equation do not exist in general. Thus, various methods were devised to investigate the seismic wavefield. The most common approaches are on the one hand based on the direct numerical solution of the wave equation, such as the finite-difference and finite-element methods, and on the other hand approximate high-frequency asymptotic methods, such as the ray method.

The aim of this chapter is to derive second-order traveltime approximations for waves propagating in inhomogeneous, isotropic 3D media by means of the ray method. For this purpose, I will introduce the eikonal equation which allows the determination of the traveltimes of waves along rays. Starting from the eikonal equation, Hamilton's equation will be derived. The latter will be used together with the concept of paraxial rays to finally end up with the sought traveltime approximations.

A very detailed description of the ray method summarising the work of many scientists who contributed to this method is provided by Červený (2001). Thus, most of the considerations I will make in Sections 2.1 and 2.2 can be found in similar form in Červený (2001). For the subsequent sections on paraxial rays, I want to refer mainly to Bortfeld (1989) and various publications by Hubral, Tygel, and Schleicher published in the 1990s, such as Hubral et al. (1992a), Hubral et al. (1992b), Tygel et al. (1992), and Schleicher et al. (1993). A very good and comprehensive summary of these papers can be found in a manuscript by Schleicher et al. (2004) which will hopefully soon be released.

2.1 The ray method

Many seismic imaging methods rely on identifying separate wave types in the recorded wavefield. Indeed, independent wave types can usually be observed in seismic records. However, as the Earth's interior is inhomogeneous, this seems to be in contradiction with the solution of the elastodynamic wave equation. The latter states that for inhomogeneous media the wavefield cannot be strictly resolved into independent waves (see, e. g., Červený, 2001). Only for homogeneous media a strict separation is possible.

The answer to this problem emerges when taking a closer look at high-frequency elastic waves. These waves approximately separate into P and S waves when propagating in smoothly inhomogeneous media. That is, both wave types do not satisfy the elastodynamic wave equation exactly but only approximately. The properties of the separated P and S wave contributions to the total wavefield are locally very similar to P and S waves propagating in homogeneous media. When the wave impinges on an interface where the medium parameters do not vary smoothly but change abruptly across the interface, reflection, transmission through the interface, as well as wave type conversion occur. In order to continue the description of wave propagation here, boundary conditions regarding the continuity of particle displacement and traction (which is related to components of the stress tensor) have to be considered across the interface.

One way to describe the propagation of high-frequency waves is the ray method. Its solution to the elastodynamic wave equation for inhomogeneous media is based on the so-called ray series. In the frequency domain, this is a series in the inverse power of the frequency ω where the series should have an asymptotic character when ω tends to infinity. Inserting the ray series into the elastodynamic wave equation leads to a non-trivial solution for all frequencies only if all coefficients of the equation obtained in this way vanish for different powers in ω. In case the frequency is sufficiently high, a set of equations can be established in this way to derive traveltimes and amplitudes of the individual waves.

The term "sufficiently high" with respect to the frequency is not a precise specification. Yet, an idea of the order of magnitude is necessary, as the applicability of the ray method depends on the prevailing frequency. In seismic literature qualitative rather than quantitative validity conditions are frequently formulated. The most common one states that the wavelength of the wave must be considerably smaller than the length of any characteristic dimension in the investigated medium.

In the following I will focus only on the kinematic aspects of high-frequency wave propagation, i. e. the calculation of traveltimes, which is governed by the so-called eikonal equation. The treatment of the dynamic part, i. e. the calculation of amplitudes, is not necessary for the comprehension of this thesis. Therefore, I will not enter into this topic. Moreover, I will restrict myself to the isotropic case where the medium is described by only two elastic parameters, the so-called Lamé parameters, as well as the density. All three parameters may vary spatially. For isotropic media, the application of high-frequency asymptotic methods leads to the same type of eikonal equation, irrespective of whether P or S waves are investigated. Thus, I will not distinguish between the individual wave types, since the considerations made in the following sections are valid for both. Additionally, for the isotropic case, the characteristics of the eikonal equation—the rays—are orthogonal trajectories to the wavefronts. This does not apply for anisotropic media which would greatly complicate the geometrical considerations made in Chapter 3. From now on, it will not be mentioned explicitly that all derivations relate to isotropic media.

2.2 The eikonal equation and the ray tracing system

The eikonal equation is a nonlinear first-order partial differential equation. In three dimensions it relates the three gradient components of the traveltime field $\tau = \tau(x, y, z)$ to the wave propagation velocity v as follows

$$\left(\hat{\nabla}\tau\right)^2 = \frac{1}{v^2},$$ (2.1)

where $\hat{\nabla} = (\partial/\partial x, \partial/\partial y, \partial/\partial z)^{\mathrm{T}}$ denotes the Nabla operator. The wave propagation velocity v may vary spatially, where the respective location is defined by vector $\hat{\mathbf{r}}$. The vectors $\hat{\nabla}$ and $\hat{\mathbf{r}} = (x, y, z)^{\mathrm{T}}$ are both given with respect to the coordinates of a global right-handed Cartesian coordinate system. The unit vectors of this system in x-, y-, and z-direction are denoted by $\hat{\mathbf{e}}_{\mathbf{x}}$, $\hat{\mathbf{e}}_{\mathbf{y}}$, and $\hat{\mathbf{e}}_{\mathbf{z}}$, respectively. A surface along which τ is constant describes a wavefront. Thus, the slowness vector $\hat{\mathbf{p}}$ which is perpendicular to the wavefront is given by

$$\hat{\mathbf{p}} = \hat{\nabla}\tau.$$ (2.2)

The vector $\hat{\mathbf{p}}$ also coincides with the wave propagation direction. Using equation (2.2), the eikonal equation (2.1) can be written in the form of a Hamilton-Jacobi equation

$$\mathcal{H}(\hat{\mathbf{r}}, \hat{\mathbf{p}}) = 0.$$ (2.3)

There are several ways to specify the Hamiltonian $\mathcal{H}$, the particular choice depending on the respective situation. For the following considerations

$$\mathcal{H} = (\hat{\mathbf{p}} \cdot \hat{\mathbf{p}})^{\frac{1}{2}} - \frac{1}{v}$$ (2.4)

is the most appropriate one.

Traditionally, the eikonal equation (2.3) is solved using the method of characteristics. The characteristic curves are called rays. They are those 3D space curves along which $\mathcal{H}(\hat{\mathbf{r}}, \hat{\mathbf{p}}) = 0$ and which are solutions to the ray tracing system:

$$\frac{d\hat{\mathbf{r}}}{ds} = v\hat{\mathbf{p}},$$ (2.5a)

$$\frac{d\hat{\mathbf{p}}}{ds} = \hat{\nabla}\left(\frac{1}{v}\right).$$ (2.5b)

The variable s denotes the monotonously increasing parameter along the ray. Its meaning is determined by the choice of the Hamiltonian $\mathcal{H}$. In the considered case s has the meaning of the arclength.

The traveltime t along the ray can be calculated independently, once the spatial trajectory is known. It is obtained by integration along the ray path from the starting point S to the end point G:

$$t = \int_S^G \frac{ds}{v} .\qquad(2.6)$$

The entire traveltime field τ is given by integration along all possible rays. Thus, for the determination of τ, the nonlinear partial differential equation (2.1) has been replaced by six ordinary differential equations (2.5) of the first order and equation (2.6). The derivation of the ray tracing system from the eikonal equation is described in detail in the book by Bleistein (1984), where also a tutorial treatment on first-order nonlinear differential equations is offered.

Three observations, which are of interest in the following, can be deduced from the ray tracing system (2.5). Firstly, the slowness vector $\hat{\mathbf{p}}$ is orthogonal to the wavefront according to equation (2.2) and, due to equation (2.5a), it is linearly dependent on the unit vector

$$\hat{\mathbf{t}} = \frac{d\hat{\mathbf{r}}}{ds} .\qquad(2.7)$$

Therefore, rays are orthogonal trajectories to the moving wavefront since $\hat{\mathbf{t}}$ is the unit tangent vector to the ray. Secondly, solving equation (2.5a) for $d\hat{\mathbf{r}}$ yields

$$d\hat{\mathbf{r}} = v\hat{\mathbf{p}}\,ds .\qquad(2.8)$$

This shows that an infinitesimally small change in the position along the ray has the same direction as the slowness vector. Finally, multiplying equation (2.8) by $\hat{\mathbf{p}}$ leads to

$$\hat{\mathbf{p}} \cdot d\hat{\mathbf{r}} = v\,(\hat{\mathbf{p}})^2\,ds = \frac{1}{v}\,ds = dt ,\qquad(2.9)$$

which reveals how an incremental step along the ray is related to an infinitesimal change dt in traveltime. All three observations will be of help later on.

2.3   Hamilton's equation

In the previous section it has been shown that the traveltimes along rays can be determined when each single ray has been traced. However, is it possible to derive—at least approximately—the traveltime along a ray from the known traveltime along a neighbouring ray?

To investigate this question, the change in traveltime due to small displacements of starting and end points of a ray is studied. For that purpose I will have a closer look at the situation depicted in Figure 2.1. Shown are two neighbouring rays which have been determined as a solution to the ray tracing system (2.5) for given initial conditions. From now on I refer to these rays as central ray and neighbouring ray. Two points on the central ray are denoted by S_0 and G_0 and on the neighbouring ray by S and G. Later on, these points will be the starting and end points of the central and neighbouring ray, respectively. Also depicted in Figure 2.1 are the wavefronts which pass through S_0 and G_0 and link the central with the neighbouring ray.

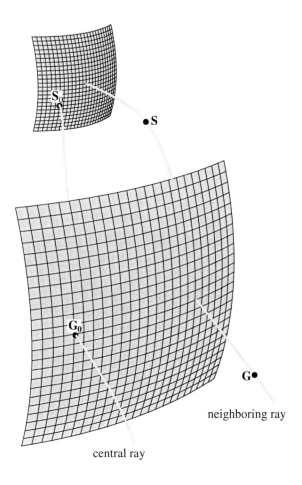

Figure 2.1: A moving wavefront at two points in time. A central and a neighbouring ray are orthogonal trajectories to this wavefront.

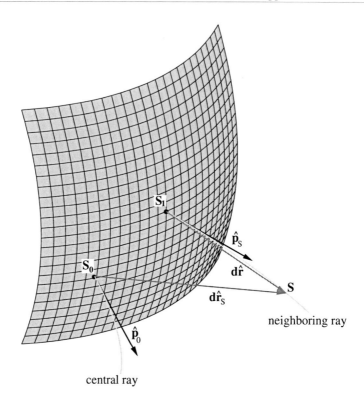

central ray

Figure 2.2: Central and neighbouring ray are orthogonal trajectories to the wavefront. Depicted are the slowness vectors $\hat{\mathbf{p}}_0$ and $\hat{\mathbf{p}}_\mathbf{S}$ at S_0 and S_1 on the wavefront. The vectors $d\hat{\mathbf{r}}$ and $d\hat{\mathbf{r}}_\mathbf{S}$ describe the dislocation from S_1 to S and S_0 to S, respectively.

An infinitesimal dislocation of a point S_1 by a vector $d\hat{\mathbf{r}}$ along the neighbouring ray to S produces a small change in the traveltime of the neighbouring ray. According to equation (2.9), this is given by

$$dt_\mathrm{S} = \hat{\mathbf{p}}_\mathbf{S} \cdot d\hat{\mathbf{r}}, \qquad (2.10)$$

where $\hat{\mathbf{p}}_\mathbf{S}$ is the slowness vector of the neighbouring ray at S_1. If S_1 is linked with S_0 through a wavefront (see Figure 2.2), the traveltime difference due to a dislocation from S_0 to S remains the same, as by definition the traveltime along a wavefront is constant. That is,

$$dt_\mathrm{S} = \hat{\mathbf{p}}_\mathbf{S} \cdot d\hat{\mathbf{r}} = \hat{\mathbf{p}}_\mathbf{S} \cdot d\hat{\mathbf{r}}_\mathbf{S}, \qquad (2.11)$$

where $d\hat{\mathbf{r}}_\mathbf{S}$ denotes the dislocation vector from S_0 to S.

The same applies for a displacement of the end position G_0. Consequently, the traveltime difference caused by the displacement from G_0 to the point G in its close vicinity is

$$dt_\mathrm{G} = \hat{\mathbf{p}}_\mathbf{G} \cdot d\hat{\mathbf{r}}_\mathbf{G}. \qquad (2.12)$$

Altogether, the infinitesimal traveltime difference dt between the central ray and the neighbouring ray reads

$$dt = dt_G - dt_S. \tag{2.13}$$

The minus sign of the second term in equation (2.13) is due to the fact that the traveltime along the neighbouring ray decreases compared to that of the central ray if the dislocation vector from S_1 to S points towards G. Inserting equations (2.11) and (2.12) into equation (2.13) yields Hamilton's equation

$$dt = \hat{\mathbf{p}}_G \cdot d\hat{\mathbf{r}}_G - \hat{\mathbf{p}}_S \cdot d\hat{\mathbf{r}}_S. \tag{2.14}$$

According to this equation a traveltime difference due to a small displacement perpendicular to the ray direction vanishes. That is, the traveltime remains stationary to infinitesimally small changes in the ray path. Therefore, Hamilton's equation is an alternative formulation of Fermat's principle. The latter states that the ray path between two points is the path that minimises the traveltime along the ray.

In fact, the ray tracing system (2.5) can be derived using Fermat's principle. The weak point of this approach, however, is the inherent assumption that P and S are two independently propagating waves even in inhomogeneous media. As stated above, this separation can be proven only by asymptotic methods.

2.4 Reflected rays

In the following chapters I will make use of reflected rays only. In particular I will work with rays, where the starting point S and end point G are located on the same plane measurement surface. This surface borders the inhomogeneous half-space with reflecting interfaces in which the different waves propagate.

For this case it is most suitable to establish the global Cartesian coordinate system in such a way that its unit vector $\hat{\mathbf{e}}_3 = (0, 0, 1)^T$ in z-direction is normal to the surface. It points away from the half-space. Thus, the z-coordinate of the system corresponds to depth.[1] The unit vectors in x- and y-direction $\hat{\mathbf{e}}_1 = (1, 0, 0)^T$ and $\hat{\mathbf{e}}_2 = (0, 1, 0)^T$, respectively, are chosen to lie in the measurement plane such that the origin of the global coordinate system also falls into this plane and the triplet $(\hat{\mathbf{e}}_1, \hat{\mathbf{e}}_2, \hat{\mathbf{e}}_3)$ is right-handed.

If the points S_0 and G_0 are the starting and end points of the central ray and S and G are those of the neighbouring ray, all of which are located in the measurement plane, the z-components of $d\hat{\mathbf{r}}_S$ and $d\hat{\mathbf{r}}_G$ vanish. For this situation Hamilton's equation (2.14) reduces to

$$dt = \mathbf{p}_G \cdot d\mathbf{r}_G - \mathbf{p}_S \cdot d\mathbf{r}_S, \tag{2.15}$$

where $\mathbf{p}_G$, $d\mathbf{r}_G$, $\mathbf{p}_S$, and $d\mathbf{r}_S$ are the projected two-component vectors in the xy-plane of the respective three-component counterpart of equation (2.14).

---

[1]The deeper a point in the medium, the smaller is its z-coordinate.

If one is interested in a second-order traveltime approximation with respect to the spatial coordinates only, Bortfeld (1989) showed that Hamilton's equation (2.14) reduces to equation (2.15) even in case of a curved measurement surface. The reason for this is that the products of the third components in equation (2.14) are already of second order and higher which would lead to terms higher than the second order in the traveltime approximation (see also Appendix B).

2.5 Paraxial rays and traveltimes

In order to compute the traveltime t along a ray from the starting point S to the end point G that resides in the vicinity of a central ray, one has to perform an integration of Hamilton's equation (2.15) with respect to $d\mathbf{r_S}$ and $d\mathbf{r_G}$. For this purpose, it is necessary to know the traveltime t_0 along the central ray and the dependence of the slowness vector projections $\mathbf{p_S}$ and $\mathbf{p_G}$, respectively, on the vectors $\mathbf{r_S}$ and $\mathbf{r_G}$. These vectors describe the dislocation from the starting and end points of the central ray to the starting and end points of the paraxial rays (see Figure 2.3). As already indicated in the previous section, a first-order approximation of $\mathbf{p_S}$ and $\mathbf{p_G}$ in $\mathbf{r_S}$ and $\mathbf{r_G}$ implies a second-order approximation of traveltime with respect to the spatial coordinates. Rays whose slowness vectors are well described by first-order approximations where the zeroth- and first-order coefficients refer to the central ray, are *paraxial* to the central ray. Consequently, the second-order traveltime approximations along paraxial rays are called *paraxial traveltimes*. The derivation of paraxial traveltimes will be discussed in the following.

To establish a linear approximation for the slowness vector projections, I will use the same notation as given in Bortfeld (1989) or Hubral et al. (1992a,b). The considerations made in Bortfeld (1989) relate only to the situation when the central ray is a normal ray, i. e., when the central ray hits the reflector normally. In this case the starting and end points as well as the two branches of the ray—from the starting point down to the reflection point and from the latter back up to the end point—are identical. I will start my derivation of paraxial traveltime approximations with the more general case where the starting and end points of the central ray do not coincide. Such a situation is depicted in Figure 2.3. Thereafter, the special case of a normal central ray will be addressed. This second case is referred to as the ZO case, recognising that coinciding starting and end points not necessarily imply that the considered ray hits the reflector normally.

2.5.1 General case

The linear approximation for $\mathbf{p_S}$ and $\mathbf{p_G}$ with respect to $\mathbf{r_S}$ and $\mathbf{r_G}$ can be formulated as follows:

$$\begin{pmatrix} \mathbf{r_G} \\ \mathbf{p_G} - \mathbf{p_{G_0}} \end{pmatrix} \approx \underline{\mathbf{T}} \begin{pmatrix} \mathbf{r_S} \\ \mathbf{p_S} - \mathbf{p_{S_0}} \end{pmatrix} \tag{2.16}$$

where

$$\underline{\mathbf{T}} = \begin{pmatrix} \mathbf{A} & \mathbf{B} \\ \mathbf{C} & \mathbf{D} \end{pmatrix}. \tag{2.17}$$

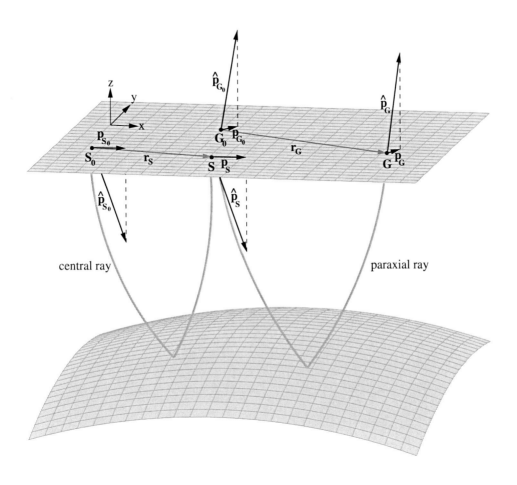

Figure 2.3: Dislocation and slowness vectors of a central and a paraxial ray at the measurement plane in the global coordinate system. S_0 and G_0 denote the starting and end points of the central ray, respectively, S and G are those of the paraxial ray.

The two-component vectors $\mathbf{p_{S_0}}$ and $\mathbf{p_{G_0}}$ are the slowness vector projections of the central ray onto the measurement plane at the starting point S_0 and the end point G_0, respectively (see Figure 2.3). The variable $\underline{\mathbf{T}}$ is a 4×4 matrix, which is commonly referred to as the surface-to-surface propagator matrix. This term is justified, since the same type of linear relationship as in equation (2.16) can be established if starting and end points are distributed on different (curved) surfaces (Bortfeld, 1989). The 2×2 submatrices $\mathbf{A}$, $\mathbf{B}$, $\mathbf{C}$, and $\mathbf{D}$ are associated with the first derivatives of the slowness vector components at S_0 and G_0. Consequently, these can be related to second derivatives of the traveltime. Solving the set of equations (2.16) for $\mathbf{p_S}$ and $\mathbf{p_G}$ yields:

$$\mathbf{p_S} \approx \mathbf{p_{S_0}} - \mathbf{B}^{-1}\mathbf{A}\,\mathbf{r_S} + \mathbf{B}^{-1}\mathbf{r_G} \tag{2.18a}$$

$$\mathbf{p_G} \approx \mathbf{p_{G_0}} + \left(\mathbf{C} - \mathbf{D}\,\mathbf{B}^{-1}\mathbf{A}\right)\mathbf{r_S} + \mathbf{D}\,\mathbf{B}^{-1}\mathbf{r_G}, \tag{2.18b}$$

where it is assumed that $\mathbf{B}^{-1}$ exists. This assumption is also made for all following considerations.

Before equations (2.18) are inserted into equation (2.15) and an integration is performed, it is necessary to have a look at the path of integration. From a physical point of view, the traveltime of the paraxial ray has to remain the same, regardless of which path of integration from the starting point of the central ray to the starting point of the paraxial ray and from the end point of the central ray to the end point of the paraxial ray is chosen. Thus, dt in equation (2.15) is a total differential. According to Schwarz' theorem this requires that the second spatial derivatives of t are independent of the order of differentiation, that is

$$\begin{pmatrix} \frac{\partial^2 t}{\partial x_S \partial x_S} & \frac{\partial^2 t}{\partial x_S \partial y_S} \\ \frac{\partial^2 t}{\partial y_S \partial x_S} & \frac{\partial^2 t}{\partial y_S \partial y_S} \end{pmatrix} = \begin{pmatrix} \frac{\partial^2 t}{\partial x_S \partial x_S} & \frac{\partial^2 t}{\partial y_S \partial x_S} \\ \frac{\partial^2 t}{\partial x_S \partial y_S} & \frac{\partial^2 t}{\partial y_S \partial y_S} \end{pmatrix}, \tag{2.19}$$

$$\begin{pmatrix} \frac{\partial^2 t}{\partial x_G \partial x_G} & \frac{\partial^2 t}{\partial x_G \partial y_G} \\ \frac{\partial^2 t}{\partial y_G \partial x_G} & \frac{\partial^2 t}{\partial y_G \partial y_G} \end{pmatrix} = \begin{pmatrix} \frac{\partial^2 t}{\partial x_G \partial x_G} & \frac{\partial^2 t}{\partial y_G \partial x_G} \\ \frac{\partial^2 t}{\partial x_G \partial y_G} & \frac{\partial^2 t}{\partial y_G \partial y_G} \end{pmatrix}, \tag{2.20}$$

and

$$\begin{pmatrix} \frac{\partial^2 t}{\partial x_S \partial x_G} & \frac{\partial^2 t}{\partial x_S \partial y_G} \\ \frac{\partial^2 t}{\partial y_S \partial x_G} & \frac{\partial^2 t}{\partial y_S \partial y_G} \end{pmatrix} = \begin{pmatrix} \frac{\partial^2 t}{\partial x_G \partial x_S} & \frac{\partial^2 t}{\partial y_G \partial x_S} \\ \frac{\partial^2 t}{\partial x_G \partial y_S} & \frac{\partial^2 t}{\partial y_G \partial y_S} \end{pmatrix}, \tag{2.21}$$

where x_S and y_S are the x- and y-components of $\mathbf{r_S}$; x_G and y_G are those of $\mathbf{r_G}$.

From equation (2.2), it is known that

$$\mathbf{p_S} = -\begin{pmatrix} \frac{\partial t}{\partial x_S} \\ \frac{\partial t}{\partial y_S} \end{pmatrix} \tag{2.22a}$$

$$\mathbf{p_G} = \begin{pmatrix} \frac{\partial t}{\partial x_S} \\ \frac{\partial t}{\partial y_S} \end{pmatrix}. \tag{2.22b}$$

The minus sign in equation (2.22a) is necessary to compensate for the minus sign in Hamilton's equation (2.15). Using equations (2.22), the first-order Taylor series for $\mathbf{p_S}$ and $\mathbf{p_G}$ can be written as follows:

$$\mathbf{p_S} \approx \mathbf{p_{S_0}} - \left(\begin{array}{cc} \frac{\partial^2 t}{\partial x_S \partial x_S} & \frac{\partial^2 t}{\partial x_S \partial y_S} \\ \frac{\partial^2 t}{\partial y_S \partial x_S} & \frac{\partial^2 t}{\partial y_S \partial y_S} \end{array} \right) \Bigg|_{x_S,y_S=0} \mathbf{r_S} - \left(\begin{array}{cc} \frac{\partial^2 t}{\partial x_S \partial x_G} & \frac{\partial^2 t}{\partial x_S \partial y_G} \\ \frac{\partial^2 t}{\partial y_S \partial x_G} & \frac{\partial^2 t}{\partial y_S \partial y_G} \end{array} \right) \Bigg|_{x_S,y_S,x_G,y_G=0} \mathbf{r_G}, \tag{2.23a}$$

$$\mathbf{p_G} \approx \mathbf{p_{G_0}} + \left(\begin{array}{cc} \frac{\partial^2 t}{\partial x_G \partial x_S} & \frac{\partial^2 t}{\partial x_G \partial y_S} \\ \frac{\partial^2 t}{\partial y_G \partial x_S} & \frac{\partial^2 t}{\partial y_G \partial y_S} \end{array} \right) \Bigg|_{x_S,y_S,x_G,y_G=0} \mathbf{r_S} + \left(\begin{array}{cc} \frac{\partial^2 t}{\partial x_G \partial x_G} & \frac{\partial^2 t}{\partial x_G \partial y_G} \\ \frac{\partial^2 t}{\partial y_G \partial x_G} & \frac{\partial^2 t}{\partial y_G \partial y_G} \end{array} \right) \Bigg|_{x_G,y_G=0} \mathbf{r_G}, \tag{2.23b}$$

where all traveltime derivatives relate to the position of the starting point and end point of the central ray. Comparing the coefficients of equations (2.23) with those of equations (2.18) provides the relationships between the submatrices of $\underline{\mathbf{T}}$ and the second derivatives of the traveltime t. These read

$$\mathbf{B}^{-1}\mathbf{A} = \left(\begin{array}{cc} \frac{\partial^2 t}{\partial x_S \partial x_S} & \frac{\partial^2 t}{\partial x_S \partial y_S} \\ \frac{\partial^2 t}{\partial y_S \partial x_S} & \frac{\partial^2 t}{\partial y_S \partial y_S} \end{array} \right) \Bigg|_{x_S,y_S=0}, \tag{2.24a}$$

$$-\mathbf{B}^{-1} = \left(\begin{array}{cc} \frac{\partial^2 t}{\partial x_S \partial x_G} & \frac{\partial^2 t}{\partial x_S \partial y_G} \\ \frac{\partial^2 t}{\partial y_S \partial x_G} & \frac{\partial^2 t}{\partial y_S \partial y_G} \end{array} \right) \Bigg|_{x_S,y_S,x_G,y_G=0}, \tag{2.24b}$$

$$\mathbf{C} - \mathbf{D}\mathbf{B}^{-1}\mathbf{A} = \left(\begin{array}{cc} \frac{\partial^2 t}{\partial x_G \partial x_S} & \frac{\partial^2 t}{\partial x_G \partial y_S} \\ \frac{\partial^2 t}{\partial y_G \partial x_S} & \frac{\partial^2 t}{\partial y_G \partial y_S} \end{array} \right) \Bigg|_{x_S,y_S,x_G,y_G=0}, \tag{2.24c}$$

$$\mathbf{D}\mathbf{B}^{-1} = \left(\begin{array}{cc} \frac{\partial^2 t}{\partial x_G \partial x_G} & \frac{\partial^2 t}{\partial x_G \partial y_G} \\ \frac{\partial^2 t}{\partial y_G \partial x_G} & \frac{\partial^2 t}{\partial y_G \partial y_G} \end{array} \right) \Bigg|_{x_G,y_G=0}. \tag{2.24d}$$

Using the symmetry of equations (2.19) and (2.20) for equations (2.24a) and (2.24d), respectively, it is obvious that

$$\mathbf{B}^{-1}\mathbf{A} = \left(\mathbf{B}^{-1}\mathbf{A} \right)^{\mathbf{T}} \tag{2.25}$$

and

$$\mathbf{D}\mathbf{B}^{-1} = \left(\mathbf{D}\mathbf{B}^{-1} \right)^{\mathbf{T}}. \tag{2.26}$$

From equations (2.24b) and (2.24c), the matrix identity

$$\mathbf{B}^{-1} = \left(\mathbf{D}\mathbf{B}^{-1}\mathbf{A} - \mathbf{C} \right)^{\mathbf{T}} \tag{2.27}$$

can be formulated, where equation (2.21) has been taken into account. This identity can be rewritten in the form

$$\mathbf{A}^{\mathbf{T}}\mathbf{D} - \mathbf{C}^{\mathbf{T}}\mathbf{B} = \mathbf{I}, \tag{2.28}$$

in which $\mathbf{I}$ denotes the 2×2 identity matrix.

The three equations (2.25), (2.26), and (2.28) together can be described by one single matrix equation defining the inverse matrix $\underline{\mathbf{T}}^{-1}$ of $\underline{\mathbf{T}}$:

$$\underline{\mathbf{T}}^{-1} = \begin{pmatrix} \mathbf{A} & \mathbf{B} \\ \mathbf{C} & \mathbf{D} \end{pmatrix}^{-1} = \begin{pmatrix} \mathbf{D}^{\mathrm{T}} & -\mathbf{B}^{\mathrm{T}} \\ -\mathbf{C}^{\mathrm{T}} & \mathbf{A}^{\mathrm{T}} \end{pmatrix}. \tag{2.29}$$

The identity $\underline{\mathbf{T}}^{-1}\underline{\mathbf{T}} = \underline{\mathbf{I}}$ then reproduces the three equations (2.25), (2.26), and (2.28), where $\underline{\mathbf{I}}$ is the 4×4 identity matrix. 4×4 matrices whose 2×2 submatrices satisfy equations (2.25), (2.26), and (2.28) are called symplectic (Červený, 2001).

Inserting equations (2.18) into equation (2.15), considering the symplecticity of $\underline{\mathbf{T}}$, and performing the integration, the desired second-order traveltime approximation for t is finally obtained:

$$t\left(\mathbf{r_S},\mathbf{r_G}\right) \approx t_0 - \mathbf{p_{S_0}}\cdot\mathbf{r_S} + \mathbf{p_{G_0}}\cdot\mathbf{r_G} + \frac{1}{2}\mathbf{r_S}\cdot\mathbf{B}^{-1}\mathbf{A}\,\mathbf{r_S} - \mathbf{r_S}\cdot\mathbf{B}^{-1}\mathbf{r_G} + \frac{1}{2}\mathbf{r_G}\cdot\mathbf{D}\mathbf{B}^{-1}\mathbf{r_G}. \tag{2.30}$$

For arbitrary starting and end point positions of a paraxial ray in the vicinity of a central ray, the associated traveltime can now be computed if the traveltime t_0 along the central ray as well as all 14 linear and quadratic coefficients are known.

2.5.2 Zero-offset case

If the central ray hits the reflector perpendicularly at the so-called normal-incidence point, which I shortly call NIP, some simplifications compared to the general case can be made. Firstly, the slowness vector projections of the central ray at starting and end point onto the measurement plane are no longer independent but related by

$$\mathbf{p_{G_0}} = -\mathbf{p_{S_0}} = \mathbf{p_0}. \tag{2.31}$$

Secondly, $\underline{\mathbf{T}}$, describing the linear approximation of position vectors and slowness vector projections, respectively, between the starting and end points for paraxial rays (see equation 2.16), can be replaced by the surface-to-surface propagator matrix $\underline{\mathbf{T}}_0$. This matrix $\underline{\mathbf{T}}_0$ is defined by

$$\underline{\mathbf{T}}_0 = \begin{pmatrix} \mathbf{A}_0 & \mathbf{B}_0 \\ \mathbf{C}_0 & \mathbf{D}_0 \end{pmatrix}, \tag{2.32}$$

where $\mathbf{A}_0$, $\mathbf{B}_0$, $\mathbf{C}_0$, and $\mathbf{D}_0$ are 2×2 matrices. These relate to second spatial derivatives of the traveltime. The matrix $\underline{\mathbf{T}}_0$ sets a linear relationship between the position vector projections and slowness vector projections for the first branch of the reflected paraxial rays (i.e. from S to R). Therefore, it is specified as a one-way propagator matrix. If the directions of the second branch of the paraxial rays are reversed, i.e., if the starting point is at G and the end point is at R, the same linear relationship using $\underline{\mathbf{T}}_0$ can be established. Consequently, linear approximations can be given in terms of $\mathbf{A}_0$, $\mathbf{B}_0$, $\mathbf{C}_0$, and $\mathbf{D}_0$ for both, $\mathbf{p_S}$ and $\mathbf{p_G}$.

In order to describe the vectors at the reflector, it is appropriate to introduce a local Cartesian coordinate system (see Figure 2.4). All coordinates in this system are primed to distinguish them

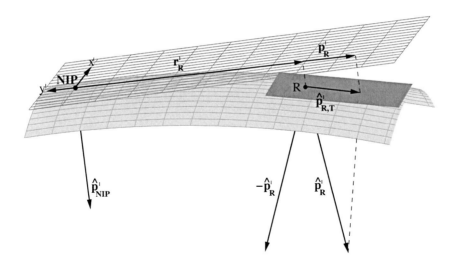

Figure 2.4: Slowness vectors and paraxial dislocation vector at the reflector. NIP denotes the reflection point of the central ray, R is the reflection point of the paraxial ray. The two-component representation of $\hat{\mathbf{p}}'_{\mathbf{R}}$ can be obtained by two cascaded projections: an orthogonal projection onto the plane which is tangent to the reflector at R followed by an orthogonal projection onto the plane which is tangent to the reflector at the NIP.

from those of the global system. Thus, I refer to the local system as the primed coordinate system. The unit vector in z'-direction points in the same direction as the slowness vector of the central ray $\hat{\mathbf{p}}'_{\mathbf{NIP}}$ at the NIP. The unit vectors in x'- and y'-direction lie in the plane which is perpendicular to $\hat{\mathbf{p}}'_{\mathbf{NIP}}$ at the NIP. The orientation of the latter two unit vectors in this plane is chosen such that the primed coordinate system is right-handed.

The use of the $4{\times}4$ matrix $\underline{\mathbf{T}}_0$ implies that the third components of the position and slowness vectors at the reflector do not need to be considered for a second-order traveltime approximation along a ray between S and R. This holds true although the reflector may be curved and the third component of the dislocation vector between the NIP and R does not vanish (as it does in case of a plane measurement surface, see Section 2.4). The proof for this fact is given in Appendix B.

The linear approximation of the slowness vector projections at S and R using $\underline{\mathbf{T}}_0$ is expressed by

$$\mathbf{p}_{\mathbf{S}} \approx -\mathbf{p}_0 - \mathbf{B}_0^{-1}\mathbf{A}_0\,\mathbf{r}_{\mathbf{S}} + \mathbf{B}_0^{-1}\mathbf{r}'_{\mathbf{R}} \tag{2.33a}$$

$$\mathbf{p}'_{\mathbf{R}} \approx \left(\mathbf{C}_0 - \mathbf{D}_0\,\mathbf{B}_0^{-1}\mathbf{A}_0\right)\mathbf{r}_{\mathbf{S}} + \mathbf{D}_0\,\mathbf{B}_0^{-1}\mathbf{r}'_{\mathbf{R}}, \tag{2.33b}$$

in which $\mathbf{r}'_{\mathbf{R}}$ is the position vector of R in the $x'y'$-plane of the primed coordinate system. The quantity $\mathbf{p}'_{\mathbf{R}}$ is the two-component vector in the $x'y'$-plane of the paraxial slowness vector at R. It is obtained by two successive projections as depicted in Figure 2.4. The first transformation projects the paraxial slowness vector onto the plane which is tangent to the reflector at R. The second transformation projects the resulting vector onto the $x'y'$-plane. Note that the two-component vector of $\mathbf{p_{NIP}}$ in the $x'y'$-plane in equation (2.33b) vanishes due to the choice of the z'-axis for the primed coordinate system.

If I reverse the direction of the second branch of the paraxial ray, i. e., if I put the starting point at G and the end point at R, as stated earlier, I can use $\underline{\mathbf{T_0}}$ to set up the paraxial approximation for the slowness vector projections:

$$-\mathbf{p_G} \approx -\mathbf{p_0} - \mathbf{B_0^{-1}A_0r_G} + \mathbf{B_0^{-1}r_R'} , \qquad (2.34a)$$

$$-\mathbf{p_R'} \approx \left(\mathbf{C_0} - \mathbf{D_0B_0^{-1}A_0}\right)\mathbf{r_G} + \mathbf{D_0B_0^{-1}r_R'} . \qquad (2.34b)$$

The minus sign on the left-hand side of equation (2.34a) reverses the direction of the slowness vector at G. The use of the slowness vector projection $-\mathbf{p_R'}$ at R in equation (2.34b), which is identical to equation (2.33b) except for the minus sign, reflects the invariance of the tangential ray slowness vectors at interfaces. Its minus sign is again necessary because of the direction reversal of the second branch of the paraxial ray. Summing equations (2.33b) and (2.34b) and solving for $\mathbf{r_R'}$, I obtain

$$\mathbf{r_R'} \approx \left(\mathbf{A_0} - \mathbf{B_0D_0^{-1}C_0}\right)\frac{1}{2}\left(\mathbf{r_G} + \mathbf{r_S}\right) . \qquad (2.35)$$

Substituting equation (2.35) into equations (2.33a) and (2.34a) finally yields $\mathbf{p_S}$ and $\mathbf{p_G}$ in terms of $\mathbf{r_S}$ and $\mathbf{r_G}$:

$$\mathbf{p_S} \approx -\mathbf{p_0} - \mathbf{B_0^{-1}A_0r_S} + \left(\mathbf{B_0^{-1}A_0} - \mathbf{D_0^{-1}C_0}\right)\frac{1}{2}\left(\mathbf{r_G} + \mathbf{r_S}\right) , \qquad (2.36a)$$

$$\mathbf{p_G} \approx \mathbf{p_0} + \mathbf{B_0^{-1}A_0r_G} - \left(\mathbf{B_0^{-1}A_0} - \mathbf{D_0^{-1}C_0}\right)\frac{1}{2}\left(\mathbf{r_G} + \mathbf{r_S}\right) . \qquad (2.36b)$$

Equations (2.36) show that linear approximations of the slowness vector projection at the starting and end point of the paraxial ray are both determined by the one-way propagator matrix $\underline{\mathbf{T_0}}$.

Inserting the slowness vector projections of equations (2.36) into Hamilton's equation (2.15) and performing an integration gives rise to the second-order traveltime approximation for paraxial rays in the vicinity of a normal ray:

$$t\left(\Delta\mathbf{m}, \mathbf{h}\right) \approx t_0 + 2\mathbf{p_0} \cdot \Delta\mathbf{m} + \Delta\mathbf{m} \cdot \mathbf{D_0^{-1}C_0}\Delta\mathbf{m} + \mathbf{h} \cdot \mathbf{B_0^{-1}A_0h} \qquad (2.37)$$

As in the general case, t_0 denotes the traveltime along the central ray. Note that t_0 is a two-way traveltime, i. e. from the starting point to the reflection point and back to the end point that coincides with the starting point. The vectors $\Delta\mathbf{m}$ and $\mathbf{h}$ explain themselves as

$$\Delta\mathbf{m} = \frac{1}{2}\left(\mathbf{r_G} + \mathbf{r_S}\right) = \mathbf{m} - \mathbf{m_0} \qquad (2.38)$$

and

$$\mathbf{h} = \frac{1}{2}\left(\mathbf{r_G} - \mathbf{r_S}\right) . \qquad (2.39)$$

The vector $\Delta\mathbf{m} = \left(\Delta m_x, \Delta m_y\right)$ denotes the dislocation vector from the midpoint between the starting and end points of the central ray (defined by vector $\mathbf{m_0} = \left(m_{x,0}, m_{y,0}\right)$) and the midpoint between the starting and end points of the paraxial ray (defined by vector $\mathbf{m}$). The vector $\mathbf{h}$ is the half-offset vector. It points from the starting point to the end point of the ray, where the norm of the vector is given by half the distance between the two points.

Let me make three remarks regarding equation (2.37). Firstly, all 2×2 matrices in equation (2.37) can be determined by one-way experiments, i. e., by rays which connect points on the measurement plane with points on the reflector. Secondly, the matrix products $\mathbf{D}_0^{-1}\mathbf{C}_0$ and $\mathbf{B}_0^{-1}\mathbf{A}_0$ lead to two symmetric 2×2 matrices. This reveals that in the ZO case there are only two independent submatrices of the propagator matrix $\underline{\mathbf{T}}$: in Appendix A it is shown that in the ZO case the submatrix $\mathbf{A}$ is related to the submatrix $\mathbf{D}$. Together with equation (2.28) this reduces the number of independent matrices in the ZO case to two. Finally, all considerations in this section are made under the assumption that both matrices $\mathbf{B}_0^{-1}$ and $\mathbf{D}_0^{-1}$ exist. Bortfeld and Kemper (1991) show situations when these matrices do not exist. They reveal that such situations occur in the case of wave focusing phenomena.

Chapter 3

Kinematic wavefield attributes

In the previous chapter it has been demonstrated that the surface-to-surface propagator matrix approach of Bortfeld (1989) offers an elegant way to formulate paraxial traveltime formulas. Moreover, Bortfeld (1989) indicated that the propagator matrix can also be used to solve various other ray-theoretical problems and is, thus, of interest in seismic applications relating seismic data to subsurface properties. However, from a geometrical point of view, the relationship between the propagator matrix and subsurface parameters is in general not obvious. In this respect, kinematic wavefield attributes are easier to interpret. These attributes describe the propagation direction and curvatures of wavefronts at the measurement surface. Therefore, kinematic wavefield attributes may directly lead to information on the reflector properties and facilitate the use of traveltime information in seismic imaging.

In this chapter I will give the mathematical foundation to describe wavefronts at the measurement surface. Thereafter, I will derive the relations between the ray slowness vectors as well as the submatrices of the propagator matrix given in equations (2.30) and (2.37) and kinematic wavefield attributes. Finally, this allows the representation of traveltime formulas in terms of these attributes.

3.1   Local ray-centred coordinate system

For the following derivations it is useful to establish a local ray-centred coordinate system at the starting or end point of the central ray on the measurement plane. To distinguish this system from the global Cartesian coordinate system (defined in Section 2.4), the coordinates, vectors, and matrices described in this system are marked with a tilde. The origin of the local ray-centred coordinate system coincides with the starting/end point of the central ray. The unit vector $\tilde{\mathbf{e}}_z$ in $\tilde{z}$-direction points in the same direction as the slowness vector of the central ray at the origin. The unit vectors $\tilde{\mathbf{e}}_x$ and $\tilde{\mathbf{e}}_y$ in $\tilde{x}$- and $\tilde{y}$-direction, respectively, fall into the plane that is perpendicular to $\tilde{\mathbf{e}}_z$ at the origin of the ray-centred coordinate system. The orientation of $\tilde{\mathbf{e}}_x$ and $\tilde{\mathbf{e}}_y$ in the $\tilde{x}\tilde{y}$-plane is not arbitrary, but chosen such that $\tilde{\mathbf{e}}_y$ lies in the intersection line of the xy-plane of the global system and the $\tilde{x}\tilde{y}$-plane and $\tilde{\mathbf{e}}_x$ makes the local ray-centred coordinate system right-handed.

3.2 Transformation from global to local ray-centred coordinates

In case the origin of the global coordinate system coincides with the origin of the local ray-centred coordinate system, the transformation between both systems can be performed by two successive rotations. For this purpose, the rotation matrices around the z- and the y-axis of the global coordinate system are required. These are

$$\hat{\mathbf{R}}_{\mathbf{z}}(\varphi) = \begin{pmatrix} \cos\varphi & -\sin\varphi & 0 \\ \sin\varphi & \cos\varphi & 0 \\ 0 & 0 & 1 \end{pmatrix} \quad \text{and} \quad \hat{\mathbf{R}}_{\mathbf{y}}(\varphi) = \begin{pmatrix} \cos\varphi & 0 & \sin\varphi \\ 0 & 1 & 0 \\ -\sin\varphi & 0 & \cos\varphi \end{pmatrix}, \tag{3.1}$$

which describe rotations with respect to the angle φ. The upper left 2×2 submatrices of $\hat{\mathbf{R}}_{\mathbf{z}}(\varphi)$ and $\hat{\mathbf{R}}_{\mathbf{y}}(\varphi)$ read

$$\mathbf{R}_{\mathbf{z}}(\varphi) = \begin{pmatrix} \cos\varphi & -\sin\varphi \\ \sin\varphi & \cos\varphi \end{pmatrix} \quad \text{and} \quad \mathbf{R}_{\mathbf{y}}(\varphi) = \begin{pmatrix} \cos\varphi & 0 \\ 0 & 1 \end{pmatrix}. \tag{3.2}$$

If the azimuth angle between $\hat{\mathbf{e}}_{\mathbf{y}}$ and $\tilde{\mathbf{e}}_{\mathbf{y}}$ is expressed by α and the polar angle between $\hat{\mathbf{e}}_{\mathbf{z}}$ and $\tilde{\mathbf{e}}_{\mathbf{z}}$ by β, then the transformation from the global to the ray-centred coordinate system is described by

$$\hat{\mathbf{R}} = \hat{\mathbf{R}}_{\mathbf{z}}(\alpha)\,\hat{\mathbf{R}}_{\mathbf{y}}(\beta), \tag{3.3}$$

where

$$-\pi < \alpha \leq \pi \tag{3.4}$$

and

$$0 \leq \beta \leq \pi/2 \tag{3.5a}$$

at the end point of the central ray and

$$\pi/2 \leq \beta \leq \pi \tag{3.5b}$$

at the starting point of the central ray. The global and a local ray-centred coordinate system as well as the meaning of the angles α and β are depicted in Figure 3.1.

Note that the matrix product of $\mathbf{R}_{\mathbf{z}}$ and $\mathbf{R}_{\mathbf{y}}$ is the upper left 2×2 submatrix $\mathbf{R}$ of $\hat{\mathbf{R}}$, that is,

$$\mathbf{R} = \mathbf{R}_{\mathbf{z}}(\alpha)\,\mathbf{R}_{\mathbf{y}}(\beta). \tag{3.6}$$

If the origin of the global coordinate system was chosen differently, an additional translation would be necessary to describe the transformation between both coordinate systems. Yet, for the transformation of relative vectors such as slowness vectors and relative position vectors (for instance, vectors describing the dislocation of the starting position of paraxial rays from those of the central ray) and for the transformation of matrices, this translation cancels out. In other words, the transformation between these vectors and matrices is still described by $\hat{\mathbf{R}}$.

For the following derivations it would be possible to arbitrarily choose the orientation of $\tilde{\mathbf{e}}_{\mathbf{x}}$ and $\tilde{\mathbf{e}}_{\mathbf{y}}$ in the $\tilde{x}\tilde{y}$-plane. However, this would require an additional rotation around the z-axis to transform the global to the ray-centred coordinates. Note that the additional rotation would not alter the orientation of the $\tilde{z}$-axis. For a detailed treatment on how to transform vectors and matrices between different coordinate systems, I refer to Höcht (2002).

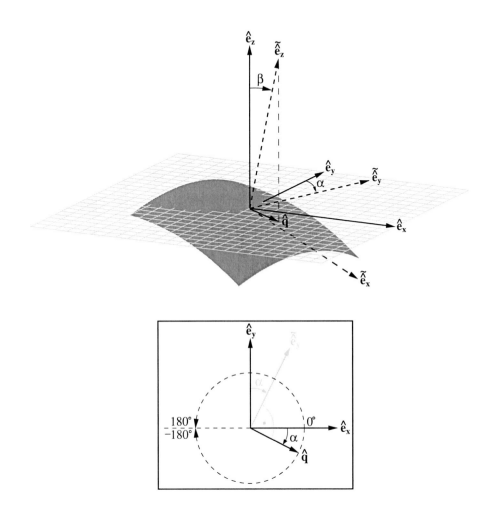

Figure 3.1: Upper part: a wavefront (shown in dark grey) emerging on the measurement surface at the end point of the central ray. The direction of the central ray at its end point determines the direction of the unit vector $\tilde{\hat{e}}_z$ in $\tilde{z}$-direction. The unit vectors $\tilde{\hat{e}}_x$ and $\tilde{\hat{e}}_y$ fall into the plane that is perpendicular to $\tilde{\hat{e}}_z$. The vector $\hat{q}$ denotes the projection of $\tilde{\hat{e}}_z$ into the xy-plane. With the knowledge of the azimuth angle α and the polar angle β, the transformation from the global coordinate system $(\hat{e}_x,\hat{e}_y,\hat{e}_z)$ to the local ray-centred coordinate system $(\tilde{\hat{e}}_x,\tilde{\hat{e}}_y,\tilde{\hat{e}}_z)$ is described. Lower part: sign convention for the azimuth angle α in the xy-plane.

3.3 Local description of a wavefront

Let the central ray and a family of paraxial rays (Section 2.5) be orthogonal trajectories to a wavefront which is starting or emerging at the starting point S_0 or the end point G_0 of the central ray. A local second-order approximation of this wavefront is described by

- the position of the starting/emerging point on the measurement plane, i. e. the position of S_0 or G_0,

- the orientation given by the propagation direction related to the first spatial derivatives of the wavefront at S_0 or G_0,

- and the curvatures of the wavefront in three different spatial directions related to the second spatial derivatives of the wavefront at S_0 or G_0.

The local ray-centred coordinate system is most suitable to present a local second-order approximation of the wavefront. There are two reasons for this. Firstly, the starting/emerging location on the measurement plane coincides with the origin of the local ray-centred coordinate system. Secondly, the first spatial derivatives of the wavefront vanish as the direction of $\tilde{\mathbf{e}}_\mathbf{z}$ coincides with the normal to the wavefront at the starting/emerging location on the measurement plane which, in turn, has the same direction as the slowness vector of the central ray at this point.

Thus, one second-order representation of the wavefront is given by

$$\tilde{z} = -\frac{1}{2}\tilde{\mathbf{r}}^\mathrm{T}\tilde{\mathbf{K}}\tilde{\mathbf{r}}\,, \tag{3.7}$$

where $\tilde{\mathbf{r}}$ is a two-component position vector in the $\tilde{x}\tilde{y}$-plane, i. e. $\tilde{\mathbf{r}} = (\tilde{x}, \tilde{y})^\mathrm{T}$. $\hat{\mathbf{K}}$ is a symmetric 2×2 curvature matrix

$$\hat{\mathbf{K}} = \begin{pmatrix} \tilde{k}_{00} & \tilde{k}_{01} \\ \tilde{k}_{01} & \tilde{k}_{11} \end{pmatrix}\,, \tag{3.8}$$

with the three independent elements $\tilde{k}_{00}$, $\tilde{k}_{01}$, and $\tilde{k}_{11}$. Equation (3.7) represents a paraboloid. However, instead of this representation, any second-order surface with identical second spatial derivatives (e. g. a hyperboloid) could be used. The minus sign in equation (3.7) is due to my sign convention of wavefront curvatures. For this convention it is necessary to take a look at the intersection of the wavefront and an arbitrarily oriented plane that is perpendicular to the tangent plane of the wavefront (i. e. the $\tilde{x}\tilde{y}$-plane). The intersection curve describes a parabola. The wavefront curvature in the intersection plane is positive if the parabola lags behind the propagating tangent plane of the wavefront. It is negative if the parabola is ahead of the propagating tangent plane (see also Figure 3.2). With this definition I follow the one given in Hubral and Krey (1980) and Höcht (2002).

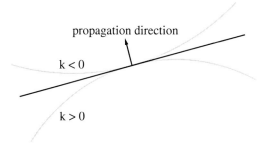

propagation direction

k < 0

k > 0

Figure 3.2: Definition of the sign of the wavefront curvature by the intersection of an arbitrarily oriented plane and the wavefront. Due to the representation of the wavefront given in equation (3.7), the intersection curve is a parabola. Two different parabolas are depicted. The parabola which lags behind the propagating tangent plane of the wavefront (indicated by the black line) has a positive curvature. The parabola which is ahead of this propagating tangent plane has a negative curvature.

3.4 Slowness vector approximation with kinematic wavefield attributes

The aim of this section is to find a first-order approximation for $\mathbf{p_S}$ and $\mathbf{p_G}$ (see Figure 2.3) in terms of kinematic wavefield attributes in order to use them to formulate traveltime equations. For this purpose, the considered subset of paraxial rays as well as the associated central ray are assumed to be orthogonal trajectories to a moving wavefront.[1] At the starting and end point of the central ray this wavefront is approximately described by the kinematic wavefield attributes which will enter into the formulation of the traveltime equations. The following derivations will refer to the end points of the rays. Equivalent considerations apply in the same way to the starting points.

In Figure 3.3 a wavefront emerging at the central point G_0 is depicted. The central ray and the paraxial rays are orthogonal trajectories to this wavefront. It is assumed that the wave propagation velocity given by v_G in the vicinity of G_0 is constant and the wavefront is locally well described by equation (3.7). Thus, the slowness vector of a paraxial ray at point P $(\tilde{x}_P, \tilde{y}_P, \tilde{z}_P)$ is defined by v_G and the unit vector $\tilde{\mathbf{n}}_\mathbf{P}$ normal to the wavefront at P. It is given by

$$\tilde{\mathbf{p}}_\mathbf{G} = \frac{1}{v_G} \tilde{\mathbf{n}}_\mathbf{P}. \qquad (3.9)$$

If the curvature matrix of the considered wavefront is given by $\tilde{\mathbf{K}}_\mathbf{G}$ which consists of the three independent elements $\tilde{k}_{G,00}$, $\tilde{k}_{G,01}$, and $\tilde{k}_{G,11}$, then $\tilde{\mathbf{n}}_\mathbf{P}$ can be expressed by

$$\tilde{\mathbf{n}}_\mathbf{P} = \frac{1}{\sqrt{(\tilde{k}_{G,00}\tilde{x}_P + \tilde{k}_{G,01}\tilde{y}_P)^2 + (\tilde{k}_{G,01}\tilde{x}_P + \tilde{k}_{G,11}\tilde{y}_P)^2 + 1}} \left(\begin{array}{c} -\tilde{\mathbf{K}}_\mathbf{G} \begin{pmatrix} \tilde{x}_P \\ \tilde{y}_P \end{pmatrix} \\ 1 \end{array} \right), \qquad (3.10)$$

---

[1]Note that all derivations are still made for isotropic media.

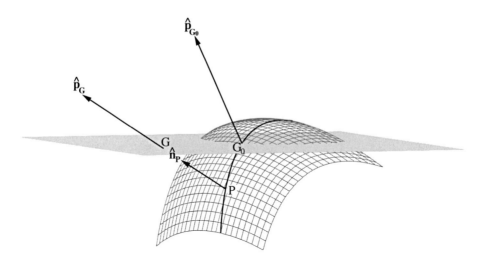

Figure 3.3: A wavefront emerging at the end point G_0 of the central ray. The central ray as well as the paraxial rays in its vicinity are orthogonal trajectories to this wavefront. The propagation direction of the wavefront is defined by the slowness vector of the central ray at G_0. If the velocity between P and G is constant, the slowness vectors of the paraxial ray passing through P and G are identical at these points.

where the wavefront description given by equation (3.7) has been taken into account.

Inserting equation (3.10) into equation (3.9) yields the slowness vector of the paraxial ray in terms of wavefront curvatures at P. Assuming v_G to be constant between P and G, the paraxial ray between both points is a straight line which is defined by the position of P and the direction vector $\tilde{\hat{\mathbf{p}}}_G$. Hence, the slowness vectors at P and G are identical. The position vector $\tilde{\mathbf{r}}_G$ of G can easily be obtained by intersecting the straight line with the measurement plane. It reads

$$\tilde{\mathbf{r}}_G = \tilde{\mathbf{r}}_P + s\,\tilde{\mathbf{n}}_P \quad \text{with} \quad s = -\hat{\mathbf{R}}_G^T \hat{\mathbf{e}}_z \cdot \tilde{\mathbf{r}}_P / \left(\hat{\mathbf{R}}_G^T \hat{\mathbf{e}}_z \cdot \tilde{\mathbf{n}}_P \right), \tag{3.11}$$

where $\tilde{\mathbf{r}}_P$ is the position vector of P. The matrix $\hat{\mathbf{R}}_G = \hat{\mathbf{R}}_z\left(\alpha_G\right) \hat{\mathbf{R}}_y\left(\beta_G\right)$ denotes the transformation from the global to the local ray-centred coordinate system at G_0 as described in Section 3.2. Note that the term $\hat{\mathbf{R}}_G^T \hat{\mathbf{e}}_z$ in equation (3.11) expresses the vector $\hat{\mathbf{e}}_z$ in the ray-centred coordinate system and is not equal to $\tilde{\hat{\mathbf{e}}}_z$.

The sought-after perpendicular projection of the paraxial slowness vector onto the measurement plane is given by

$$\tilde{\hat{\mathbf{p}}}_{G,ms} = \tilde{\hat{\mathbf{p}}}_G - \tilde{\hat{\mathbf{p}}}_{G,\perp}, \tag{3.12}$$

where $\tilde{\hat{\mathbf{p}}}_{G,\perp}$ is the component of $\tilde{\hat{\mathbf{p}}}_G$ perpendicular to the measurement plane which is expressed by

$$\tilde{\hat{\mathbf{p}}}_{G,\perp} = \hat{\mathbf{R}}_{zy}^T \hat{\mathbf{e}}_z \left(\hat{\mathbf{R}}_{zy}^T \hat{\mathbf{e}}_z \cdot \tilde{\hat{\mathbf{p}}}_G \right). \tag{3.13}$$

Considering equations (3.9) – (3.13), one can find that $\tilde{\hat{\mathbf{p}}}_{\mathrm{G,ms}}$ is a function of $\tilde{x}_{\mathrm{P}}$ and $\tilde{y}_{\mathrm{P}}$ which are, in turn, functions of $\tilde{x}_{\mathrm{G}}$ and $\tilde{y}_{\mathrm{G}}$, i. e.

$$\tilde{\hat{\mathbf{p}}}_{\mathrm{G,ms}} = \begin{pmatrix} f_1(\tilde{x}_{\mathrm{P}}(\tilde{x}_{\mathrm{G}},\tilde{y}_{\mathrm{G}}),\tilde{y}_{\mathrm{P}}(\tilde{x}_{\mathrm{G}},\tilde{y}_{\mathrm{G}})) \\ f_2(\tilde{x}_{\mathrm{P}}(\tilde{x}_{\mathrm{G}},\tilde{y}_{\mathrm{G}}),\tilde{y}_{\mathrm{P}}(\tilde{x}_{\mathrm{G}},\tilde{y}_{\mathrm{G}})) \\ f_3(\tilde{x}_{\mathrm{P}}(\tilde{x}_{\mathrm{G}},\tilde{y}_{\mathrm{G}}),\tilde{y}_{\mathrm{P}}(\tilde{x}_{\mathrm{G}},\tilde{y}_{\mathrm{G}})) \end{pmatrix} . \tag{3.14}$$

Unfortunately, there are no simple solutions for $\tilde{x}_{\mathrm{P}}$ and $\tilde{y}_{\mathrm{P}}$ in terms of $\tilde{x}_{\mathrm{G}}$ and $\tilde{y}_{\mathrm{G}}$ which becomes obvious from equation (3.11). Therefore, and because I am looking for a first-order approximation of $\tilde{\hat{\mathbf{p}}}_{\mathrm{G,ms}}$ as function of $\tilde{x}_{\mathrm{G}}$ and $\tilde{y}_{\mathrm{G}}$, equation (3.14) is expanded in a Taylor series up to the first order in $\tilde{x}_{\mathrm{G}}$ and $\tilde{y}_{\mathrm{G}}$ at the point $(\tilde{x}_{\mathrm{G}} = 0, \tilde{y}_{\mathrm{G}}, \tilde{z}_{\mathrm{G}} = 0)$. This reads

$$\tilde{\hat{\mathbf{p}}}_{\mathrm{G,ms}} \approx \tilde{\hat{\mathbf{p}}}_{\mathrm{G_0,ms}} + \begin{pmatrix} \frac{\partial f_1}{\partial \tilde{x}_{\mathrm{P}}}\tilde{x}_{\mathrm{G}} + \frac{\partial f_1}{\partial \tilde{y}_{\mathrm{P}}}\tilde{y}_{\mathrm{G}} \\ \frac{\partial f_2}{\partial \tilde{x}_{\mathrm{P}}}\tilde{x}_{\mathrm{G}} + \frac{\partial f_2}{\partial \tilde{y}_{\mathrm{P}}}\tilde{y}_{\mathrm{G}} \\ \frac{\partial f_3}{\partial \tilde{x}_{\mathrm{P}}}\tilde{x}_{\mathrm{G}} + \frac{\partial f_3}{\partial \tilde{y}_{\mathrm{P}}}\tilde{y}_{\mathrm{G}} \end{pmatrix} , \tag{3.15}$$

where $\tilde{\hat{\mathbf{p}}}_{\mathrm{G_0,ms}}$ is the perpendicular projection of the central slowness vector onto the measurement surface. It is given by

$$\tilde{\hat{\mathbf{p}}}_{\mathrm{G_0,ms}} = \frac{1}{v_{\mathrm{G}}} \begin{pmatrix} \cos\beta_{\mathrm{G}}\sin\beta_{\mathrm{G}} \\ 0 \\ \sin^2\beta_{\mathrm{G}} \end{pmatrix} . \tag{3.16}$$

In equation (3.15) the results of the four derivatives $\frac{\partial \tilde{x}_{\mathrm{P}}}{\partial \tilde{x}_{\mathrm{G}}}$, $\frac{\partial \tilde{y}_{\mathrm{P}}}{\partial \tilde{x}_{\mathrm{G}}}$, $\frac{\partial \tilde{x}_{\mathrm{P}}}{\partial \tilde{y}_{\mathrm{G}}}$, and $\frac{\partial \tilde{y}_{\mathrm{P}}}{\partial \tilde{y}_{\mathrm{G}}}$ at the point $(\tilde{x}_{\mathrm{G}} = 0, \tilde{y}_{\mathrm{G}}, \tilde{z}_{\mathrm{G}} = 0)$ have already been incorporated. The derivatives can be summarised in the following matrix equation:

$$\begin{pmatrix} \left.\frac{\partial \tilde{x}_{\mathrm{P}}}{\partial \tilde{x}_{\mathrm{G}}}\right|_{(\tilde{x}_{\mathrm{G}},\tilde{y}_{\mathrm{G}},\tilde{z}_{\mathrm{G}}=0)} & \left.\frac{\partial \tilde{x}_{\mathrm{P}}}{\partial \tilde{y}_{\mathrm{G}}}\right|_{(\tilde{x}_{\mathrm{G}},\tilde{y}_{\mathrm{G}},\tilde{z}_{\mathrm{G}}=0)} \\ \left.\frac{\partial \tilde{y}_{\mathrm{P}}}{\partial \tilde{x}_{\mathrm{G}}}\right|_{(\tilde{x}_{\mathrm{G}},\tilde{y}_{\mathrm{G}},\tilde{z}_{\mathrm{G}}=0)} & \left.\frac{\partial \tilde{y}_{\mathrm{P}}}{\partial \tilde{y}_{\mathrm{G}}}\right|_{(\tilde{x}_{\mathrm{G}},\tilde{y}_{\mathrm{G}},\tilde{z}_{\mathrm{G}}=0)} \end{pmatrix} = \hat{\mathbf{I}} , \tag{3.17}$$

where $\hat{\mathbf{I}}$ is the 2×2 identity matrix. Calculating the remaining six simple derivatives in equation (3.15) yields

$$\tilde{\hat{\mathbf{p}}}_{\mathrm{G,ms}} \approx \tilde{\hat{\mathbf{p}}}_{\mathrm{G_0,ms}} + \frac{1}{v_{\mathrm{G}}} \begin{pmatrix} \tilde{k}_{\mathrm{G,00}}\cos^2\beta_{\mathrm{G}} & \tilde{k}_{\mathrm{G,01}}\cos^2\beta_{\mathrm{G}} & 0 \\ \tilde{k}_{\mathrm{G,01}} & \tilde{k}_{\mathrm{G,11}} & 0 \\ \tilde{k}_{\mathrm{G,00}}\sin\beta_{\mathrm{G}}\cos\beta_{\mathrm{G}} & \tilde{k}_{\mathrm{G,01}}\sin\beta_{\mathrm{G}}\cos\beta_{\mathrm{G}} & 0 \end{pmatrix} \tilde{\mathbf{r}}_{\mathrm{G}} . \tag{3.18}$$

Transforming $\tilde{\hat{\mathbf{p}}}_{\mathrm{G,ms}}$ in equation (3.18) into the global coordinate system leads to

$$\hat{\mathbf{p}}_{\mathrm{G,ms}} = \hat{\mathbf{R}}_{\mathrm{G}}\tilde{\hat{\mathbf{p}}}_{\mathrm{G,ms}} \approx \frac{1}{v_{\mathrm{G}}} \begin{pmatrix} \cos\alpha_{\mathrm{G}}\sin\beta_{\mathrm{G}} \\ \sin\alpha_{\mathrm{G}}\sin\beta_{\mathrm{G}} \\ 0 \end{pmatrix} + \frac{1}{v_{\mathrm{G}}} \begin{pmatrix} \mathbf{R}_{\mathrm{G}}\tilde{\mathbf{K}}_{\mathrm{G}}\mathbf{R}_{\mathrm{G}}^{\mathrm{T}} \\ 0 \end{pmatrix} \hat{\mathbf{r}}_{\mathrm{G}} . \tag{3.19}$$

The vector $\left(\cos\alpha_G\sin\beta_G, \sin\alpha_G\sin\beta_G\right)^T$ represents the first two components of $\hat{\mathbf{p}}_{G_0}$ since

$$\hat{\mathbf{p}}_{G_0} = \frac{1}{v_G}\hat{\mathbf{R}}_G\tilde{\mathbf{e}}_z = \frac{1}{v_G}\begin{pmatrix} \cos\alpha_G\sin\beta_G \\ \sin\alpha_G\sin\beta_G \\ \cos\beta_G \end{pmatrix}. \tag{3.20}$$

Thus, the first two components of $\hat{\mathbf{p}}_{G,\text{ms}}$ can be written as

$$\mathbf{p}_G \approx \mathbf{p}_{G_0} + \frac{1}{v_G}\mathbf{R}_G\tilde{\mathbf{K}}_G\mathbf{R}_G^T\mathbf{r}_G, \tag{3.21}$$

which is the sought-after approximation of the slowness vector projection.

The same line of argument is valid for the starting points of central and paraxial rays. These are assumed to be orthogonal trajectories to a wavefront with the wavefront curvature matrix $\tilde{\mathbf{K}}_S$ at the starting point S_0 of the central ray. Therefore, the slowness vector projections of the considered paraxial rays are approximated by

$$\mathbf{p}_S \approx \mathbf{p}_{S_0} + \frac{1}{v_S}\mathbf{R}_S\tilde{\mathbf{K}}_S\mathbf{R}_S^T\mathbf{r}_S, \tag{3.22}$$

where v_S denotes the wave propagation velocity in the vicinity of S_0. The transformation matrix $\mathbf{R}_S$ is the result of the matrix product $\hat{\mathbf{R}}_z\left(\alpha_S\right)\hat{\mathbf{R}}_y\left(\beta_S\right)$. The meaning of α_S as well as β_S is again as explained in Section 3.2. The vector $\mathbf{p}_{S_0}$ denotes the first two components of the vector $\hat{\mathbf{p}}_{S_0}$ which is given by

$$\hat{\mathbf{p}}_{S_0} = \frac{1}{v_S}\begin{pmatrix} \cos\alpha_S\sin\beta_S \\ \sin\alpha_S\sin\beta_S \\ \cos\beta_S \end{pmatrix}. \tag{3.23}$$

Note that equations (2.18) represent the paraxial slowness vectors for all paraxial rays with different starting and end positions. Thus, their first-order traveltime coefficients describe the full surface-to-surface propagator matrix. Equations (3.21) and (3.22) are valid for a subset of paraxial rays, namely those rays which belong to a specific wavefront. Different experiments resulting in different wavefronts at the starting point and end point of the central ray allow the determination of the full surface-to-surface propagator matrix. These experiments are introduced in the following section and used to formulate the paraxial traveltime in terms of kinematic wavefront attributes.

3.5 Traveltime formulas with kinematic wavefield attributes

In this section I relate the submatrices of the surface-to-surface propagator matrix $\underline{\mathbf{T}}$ to kinematic wavefield attributes which are the result of different experiments. In this way, the first- and second-order coefficients gain a geometrical meaning which makes their interpretation easier. As in Section 2.5, I will divide this section into the general case (where the starting and end position of the central ray are not identical) and the ZO case (where the starting and end position of the central ray coincide).

3.5.1 General case

In Section 2.5 it has been shown that the results of the matrix products $\mathbf{B}^{-1}\mathbf{A}$ and $\mathbf{D}\mathbf{B}^{-1}$ in equation (2.30) are symmetric but not the matrix $\mathbf{B}^{-1}$ itself. As discussed in Schleicher (1993), $\mathbf{B}^{-1}$ is only symmetric if the central ray has no torsion. This occurs, for instance, for media which do not vary in one spatial direction (the so-called 2.5D case). In general, the total number of independent second-order coefficients in equation (2.30) is therefore ten.

A curvature matrix as given in equation (3.8) has three independent elements. Thus, the knowledge of at least four different wavefronts is necessary if all ten second-order traveltime coefficients are to be related to wavefront curvatures. For this purpose, I introduce four experiments. Besides these, many other experiments could be chosen instead. Only for the determination of $\mathbf{B}^{-1}$ it will be demonstrated that the experiments have to fulfil certain properties.

Common-shot experiment

The first selected experiment is the common-shot (CS) experiment. This is performed by placing a point source at the starting point of the central ray. Then, the wavefront of the initiated wave propagates along the rays involved in this experiment and is registered at their end points where receivers are located. The initial condition for all rays involved in this experiment (CS rays) is expressed by

$$\mathbf{r_S} = \begin{pmatrix} 0 \\ 0 \end{pmatrix}. \tag{3.24}$$

The traveltime approximation along the CS rays located in the vicinity of the central ray is obtained by substituting equation (3.24) into equation (2.30) which yields:

$$t_{CS}\left(\mathbf{r_G}\right) \approx t_0 + \mathbf{p_{G_0}} \cdot \mathbf{r_G} + \frac{1}{2}\mathbf{r_G} \cdot \mathbf{D}\mathbf{B}^{-1}\mathbf{r_G}. \tag{3.25}$$

Thus, the first-order approximation of the slowness vector projection $\mathbf{p_G^{CS}}$ of the CS rays at the end points is given by

$$\frac{d t_{CS}}{d\mathbf{r_G}} = \begin{pmatrix} \frac{\partial t_{CS}}{\partial x_G} \\ \frac{\partial t_{CS}}{\partial y_G} \end{pmatrix} = \mathbf{p_G^{CS}} \approx \mathbf{p_{G_0}} + \mathbf{D}\mathbf{B}^{-1}\mathbf{r_G}. \tag{3.26}$$

Let the curvatures of the propagating wavefronts at the end point of the central ray be described by the matrix $\tilde{\mathbf{K}}_{\mathbf{G}}^{\mathbf{CS}}$. According to equation (3.21), $\mathbf{p_G^{CS}}$ can then be approximated in terms of the curvature matrix by

$$\mathbf{p_G^{CS}} \approx \mathbf{p_{G_0}} + \frac{1}{v_G}\mathbf{R_G}\tilde{\mathbf{K}}_{\mathbf{G}}^{\mathbf{CS}}\mathbf{R_G^T}\mathbf{r_G}. \tag{3.27}$$

Comparing the coefficients of equations (3.26) and (3.27) yields

$$\mathbf{D}\mathbf{B}^{-1} = \frac{1}{v_G}\mathbf{R}_G\tilde{\mathbf{K}}_G^{CS}\mathbf{R}_G^T. \tag{3.28}$$

Therefore, the second traveltime derivatives with respect to the coordinates of $\mathbf{r}_G$ are related to wavefront curvatures, when taking equation (2.24d) into account.

Common-receiver experiment

The second considered experiment is the common-receiver (CR) experiment. For this, a wave is initiated which focuses at the end point of the central ray. The rays associated with the CR experiment (CR rays) are identical to those involved in an experiment, where a point source is placed at the end point of the central ray and the propagating wavefront of the initiated wave is detected at receivers located at the starting points of the CR rays. Consequently, the initial condition of the CR experiment is given by

$$\mathbf{r}_G = \begin{pmatrix} 0 \\ 0 \end{pmatrix}. \tag{3.29}$$

Inserting equation (3.29) into equation (2.30) yields the traveltime approximation along the CR rays located in the vicinity of the central ray:

$$t_{CR}(\mathbf{r}_S) \approx t_0 - \mathbf{p}_{S_0}\cdot\mathbf{r}_S + \frac{1}{2}\mathbf{r}_S\cdot\mathbf{B}^{-1}\mathbf{A}\,\mathbf{r}_S. \tag{3.30}$$

Taking the first derivatives of t_{CR} with respect to the components of $\mathbf{r}_S$, it follows for the slowness vector projections $\mathbf{p}_S^{CR}$ of the CR rays at their starting points that

$$-\frac{dt_{CR}}{d\mathbf{r}_S} = -\begin{pmatrix} \frac{\partial t_{CR}}{\partial x_S} \\ \frac{\partial t_{CR}}{\partial y_S} \end{pmatrix} = \mathbf{p}_S^{CR} \approx \mathbf{p}_{S_0} - \mathbf{B}^{-1}\mathbf{A}\,\mathbf{r}_S. \tag{3.31}$$

As already discussed in Chapter 2, the minus sign on the left-hand side of equation (3.31) is due the minus sign in equation (2.15).

If the curvatures of the propagating wavefront at the starting point of the central ray is described by $\tilde{\mathbf{K}}_S$, then $\mathbf{p}_S^{CR}$ can, according to equation (3.21), be approximated by

$$\mathbf{p}_S^{CR} \approx \mathbf{p}_{S_0} + \frac{1}{v_S}\mathbf{R}_S\tilde{\mathbf{K}}_S^{CR}\mathbf{R}_S^T\,\mathbf{r}_S. \tag{3.32}$$

Comparing the coefficients of equations (3.31) and (3.32), one can find that

$$\mathbf{B}^{-1}\mathbf{A} = -\frac{1}{v_S}\mathbf{R}_S\tilde{\mathbf{K}}_S^{CR}\mathbf{R}_S^T. \tag{3.33}$$

Hence, equation (3.33) relates, according to equation (2.24a), the second traveltime derivatives with respect to the coordinates of $\mathbf{r}_S$ to wavefront curvatures.

Common-midpoint experiment

The mixed second traveltime derivatives represented by matrix $\mathbf{B}^{-1}$ (see equation (2.24b)) are still to be related to wavefront curvatures. Since $\mathbf{B}^{-1}$ is non-symmetric, two different experiments are necessary for this purpose. The first is the common-midpoint (CMP) experiment. The starting and end points of all rays involved in the CMP experiment (CMP rays) are subject to the condition

$$\mathbf{r_G} = -\mathbf{r_S} = \Delta\mathbf{h} = \begin{pmatrix} \Delta h_x \\ \Delta h_y \end{pmatrix}. \tag{3.34}$$

The vector $\Delta\mathbf{h}$ describes the dislocation between the half-offset vector $\mathbf{h}$ of the paraxial rays and the half-offset vector $\mathbf{h_0}$ of the central ray. Substituting equation (3.34) into equation (2.30) gives rise to the CMP traveltime formula

$$\begin{aligned} t_{\text{CMP}}(\Delta\mathbf{h}) &\approx t_0 + \left(\mathbf{p_{S_0}} + \mathbf{p_{G_0}} \right) \cdot \Delta\mathbf{h} + \Delta\mathbf{h} \cdot \left(\frac{1}{2}\mathbf{B}^{-1}\mathbf{A} + \mathbf{B}^{-1} + \frac{1}{2}\mathbf{D}\mathbf{B}^{-1} \right) \Delta\mathbf{h} \\ &= t_0 + \left(\mathbf{p_{S_0}} + \mathbf{p_{G_0}} \right) \cdot \Delta\mathbf{h} + \frac{1}{2}\Delta\mathbf{h} \cdot \left(\mathbf{B}^{-1}\mathbf{A} + \mathbf{B}^{-1} + \mathbf{B}^{-T} + \mathbf{D}\mathbf{B}^{-1} \right) \Delta\mathbf{h}. \end{aligned} \tag{3.35}$$

The rearrangement in equation (3.35) is possible as the mixed second derivatives $\partial^2 t_{\text{CMP}}/\partial\Delta h_x \partial\Delta h_y$ and $\partial^2 t_{\text{CMP}}/\partial\Delta h_y \partial\Delta h_x$ are identical. Note that equation (3.35) does not allow the conclusion that $2\mathbf{B}^{-1} = \mathbf{B}^{-1} + \mathbf{B}^{-T}$. Taking the first derivatives of t_{CMP} with respect to Δh_x and Δh_y yields

$$\frac{dt_{\text{CMP}}}{d(\Delta\mathbf{h})} = \begin{pmatrix} \frac{\partial t_{\text{CMP}}}{\partial(\Delta h_x)} \\ \frac{\partial t_{\text{CMP}}}{\partial(\Delta h_y)} \end{pmatrix} \approx \mathbf{p_{S_0}} + \mathbf{p_{G_0}} + \left(\mathbf{B}^{-1}\mathbf{A} + \mathbf{B}^{-1} + \mathbf{B}^{-T} + \mathbf{D}\mathbf{B}^{-1} \right)\Delta\mathbf{h}. \tag{3.36}$$

Using equation (3.34) and the chain rule for derivatives to calculate $\frac{dt_{\text{CMP}}}{d(\Delta\mathbf{h})}$, one can separate the influence of the slowness vector projection at the starting points and end points of the CMP rays, respectively, on $\frac{dt_{\text{CMP}}}{d(\Delta\mathbf{h})}$:

$$\frac{dt_{\text{CMP}}}{d(\Delta\mathbf{h})} = \begin{pmatrix} \frac{\partial t_{\text{CMP}}}{\partial x_S}\frac{dx_S}{d(\Delta h_x)} + \frac{\partial t_{\text{CMP}}}{\partial x_G}\frac{dx_G}{d(\Delta h_x)} \\ \frac{\partial t_{\text{CMP}}}{\partial y_S}\frac{dy_S}{d(\Delta h_y)} + \frac{\partial t_{\text{CMP}}}{\partial y_G}\frac{dy_G}{d(\Delta h_y)} \end{pmatrix} = \begin{pmatrix} -\frac{\partial t_{\text{CMP}}}{\partial x_S} + \frac{\partial t_{\text{CMP}}}{\partial x_G} \\ -\frac{\partial t_{\text{CMP}}}{\partial y_S} + \frac{\partial t_{\text{CMP}}}{\partial y_G} \end{pmatrix} = \mathbf{p_S^{CMP}} + \mathbf{p_G^{CMP}}, \tag{3.37}$$

where $\mathbf{p_S^{CMP}}$ denotes the slowness vector projections of the CMP rays at their starting points and $\mathbf{p_G^{CMP}}$ denotes the slowness vector projections of the CMP rays at their end points. With equations (3.21) and (3.22) $\frac{dt_{\text{CMP}}}{d(\Delta\mathbf{h})}$ can now be related to the wavefront curvatures of the CMP experiment:

$$\frac{dt_{\text{CMP}}}{d(\Delta\mathbf{h})} \approx \mathbf{p_{S_0}} + \mathbf{p_{G_0}} + \left(-\frac{1}{v_S}\mathbf{R_S}\tilde{\mathbf{K}}_S^{\mathbf{CMP}}\mathbf{R_S^T} + \frac{1}{v_G}\mathbf{R_G}\tilde{\mathbf{K}}_G^{\mathbf{CMP}}\mathbf{R_G^T} \right)\Delta\mathbf{h}. \tag{3.38}$$

The matrices $\tilde{\mathbf{K}}_S^{\mathbf{CMP}}$ and $\tilde{\mathbf{K}}_G^{\mathbf{CMP}}$ describe the curvatures of a propagating wavefront at the starting and end points of the central ray, where the wavefront is an orthogonal surface to the CMP rays. It

33

is impossible to initiate a wave, to which this wavefront belongs, by means of a single experiment. In fact, the CMP experiment consists of many single experiments, where the CMP wavefront can be looked upon as an envelope of many wavefronts. Therefore, the CMP wave is denoted as a hypothetical wave.

Comparing the coefficients of equations (3.36) and (3.38) shows that

$$\mathbf{B}^{-1} + \mathbf{B}^{-T} = \frac{1}{v_S} \mathbf{R}_S \left(\tilde{\mathbf{K}}_S^{CR} - \tilde{\mathbf{K}}_S^{CMP} \right) \mathbf{R}_S^T + \frac{1}{v_G} \mathbf{R}_G \left(\tilde{\mathbf{K}}_G^{CMP} - \tilde{\mathbf{K}}_G^{CS} \right) \mathbf{R}_G^T, \qquad (3.39)$$

where equations (3.28) and (3.33)—the results of the CS and CR experiments—have been taken into account. If the inverse of matrix $\mathbf{B}$ is denoted by the matrix $\mathbf{E}$, which is defined by

$$\mathbf{B}^{-1} = \mathbf{E} = \begin{pmatrix} e_{00} & e_{01} \\ e_{10} & e_{11} \end{pmatrix}, \qquad (3.40)$$

I find that it is possible to calculate the elements e_{00} and e_{11} by the matrix equation (3.39). However, it is impossible to obtain independent solutions for the non-diagonal elements e_{01} and e_{10} but only a solution for the sum $e_{01} + e_{10}$. This is due to the fact that the matrices on both sides of equation (3.39), resulting from the different products and sums, are symmetric. Moreover, independent solutions for the non-diagonal elements of $\mathbf{E}$ could not be obtained even if I would perform additional experiments which are defined by the condition

$$\mathbf{r}_G = l\mathbf{r}_S, \qquad (3.41)$$

where l is some arbitrarily chosen real value. For all those experiments, again, only a solution for $e_{01} + e_{10}$ is obtained.

Cross-profile experiment

In order to determine the non-diagonal elements of the inverse matrix $\mathbf{B}$ (equation (3.40)), Tygel et al. (1992) and Schleicher (1993) proposed a cross-profile (CP) experiment. The starting and end points of all rays involved in the CP experiment (CP rays) have to fulfil the conditions

$$\mathbf{r}_S = \mathbf{k} \quad \text{and} \quad \mathbf{r}_G = \mathbf{Q}\mathbf{k}, \qquad (3.42)$$

where

$$\mathbf{k} = \begin{pmatrix} k_x \\ k_y \end{pmatrix} \qquad (3.43)$$

and

$$\mathbf{Q} = \begin{pmatrix} 0 & -1 \\ 1 & 0 \end{pmatrix}. \qquad (3.44)$$

This means that in a cross-profile experiment the x_G-coordinate is linearly dependent on the y_S-coordinate, and the y_G-coordinate is linearly dependent on the x_S-coordinate. Substituting equations (3.42) into equation (2.30) yields the CP traveltime formula

$$t(\mathbf{k}) \approx t_0 + \left(-\mathbf{p}_{S_0}^T + \mathbf{p}_{G_0}^T \mathbf{Q} \right) \mathbf{k} + \frac{1}{2} \mathbf{k} \cdot \left(\mathbf{B}^{-1}\mathbf{A} - \mathbf{B}^{-1}\mathbf{Q} - \mathbf{Q}^T\mathbf{B}^{-T} + \mathbf{Q}^T\mathbf{D}\mathbf{B}^{-1}\mathbf{Q} \right) \mathbf{k}. \quad (3.45)$$

Taking the first derivatives of t_{CP} with respect to k_x and k_y yields

$$\frac{dt_{CP}}{d\mathbf{k}} = \begin{pmatrix} \frac{\partial t_{CP}}{\partial k_x} \\ \frac{\partial t_{CP}}{\partial k_y} \end{pmatrix} \approx -\mathbf{p}_{S_0} + \mathbf{Q}^T\mathbf{p}_{G_0} + \left(\mathbf{B}^{-1}\mathbf{A} - \mathbf{B}^{-1}\mathbf{Q} - \mathbf{Q}^T\mathbf{B}^{-T} + \mathbf{Q}^T\mathbf{D}\mathbf{B}^{-1}\mathbf{Q} \right) \mathbf{k}. \quad (3.46)$$

Analogous to the CMP experiment, it is possible to use equation (3.42) and the chain rule for derivatives to separate the influence of the slowness vector projection at the starting points and end points of the CP rays on $\frac{dt_{CP}}{d\mathbf{k}}$:

$$\frac{dt_{CP}}{d\mathbf{k}} = \begin{pmatrix} \frac{\partial t_{CP}}{\partial x_S}\frac{dx_S}{dk_x} + \frac{\partial t_{CP}}{\partial y_G}\frac{dy_G}{dk_x} \\ \frac{\partial t_{CP}}{\partial y_S}\frac{dy_S}{dk_y} + \frac{\partial t_{CP}}{\partial x_G}\frac{dx_G}{dk_y} \end{pmatrix} = \begin{pmatrix} \frac{\partial t_{CP}}{\partial x_S} + \frac{\partial t_{CP}}{\partial y_G} \\ \frac{\partial t_{CP}}{\partial y_S} - \frac{\partial t_{CP}}{\partial x_G} \end{pmatrix} = -\mathbf{p}_S^{CP} + \mathbf{Q}^T\mathbf{p}_G^{CP}. \quad (3.47)$$

Taking equations (3.21) and (3.22) into account, $\frac{dt_{CP}}{d\mathbf{k}}$ can be related to the wavefront curvatures of the CP experiment:

$$\frac{dt_{CP}}{d\mathbf{k}} \approx -\mathbf{p}_{S_0} + \mathbf{Q}^T\mathbf{p}_{G_0} + \left(-\frac{1}{v_S}\mathbf{R}_S\tilde{\mathbf{K}}_S^{CP}\mathbf{R}_S^T + \frac{1}{v_G}\mathbf{Q}^T\mathbf{R}_G\tilde{\mathbf{K}}_G^{CP}\mathbf{R}_G^T\mathbf{Q} \right) \mathbf{k}. \quad (3.48)$$

The matrices $\tilde{\mathbf{K}}_S^{CP}$ and $\tilde{\mathbf{K}}_G^{CP}$ describe the curvatures of a propagating wavefront at the starting and end points of the central ray which is a surface orthogonal to the CP rays. As for the CMP experiment, this wavefront belongs to a hypothetical wave as the wavefront can only be constructed by superposition of many single experiments.

Comparing the coefficients of equations (3.46) and (3.48) leads to

$$\mathbf{B}^{-1}\mathbf{Q} + \mathbf{Q}^T\mathbf{B}^{-T} = \frac{1}{v_S}\mathbf{R}_S\left(\tilde{\mathbf{K}}_S^{CP} - \tilde{\mathbf{K}}_S^{CR} \right)\mathbf{R}_S^T + \frac{1}{v_G}\mathbf{Q}^T\mathbf{R}_G\left(\tilde{\mathbf{K}}_G^{CS} - \tilde{\mathbf{K}}_G^{CP} \right)\mathbf{R}_G^T\mathbf{Q}, \quad (3.49)$$

where equations (3.28) and (3.33) have already been considered. Identifying again the inverse of matrix $\mathbf{B}$ by matrix $\mathbf{E}$, it is, as a consequence of equation (3.49), possible to derive the non-diagonal elements e_{01} and e_{10} by means of wavefront curvatures. However, it is impossible to calculate solutions for the diagonal elements e_{00} and e_{11}. Similar to the experiments characterised by equation (3.41) which do not allow to compute the non-diagonal elements, all experiments characterised by

$$\mathbf{r}_G = l\mathbf{Q}\mathbf{r}_S \quad (3.50)$$

do not allow the calculation of the diagonal elements of $\mathbf{E}$, but only the difference $e_{00} - e_{11}$.

As a result of the CMP and CP experiment, the mixed second derivatives of the traveltime are related to wavefront curvatures when taking equations (2.24b), (3.39), and (3.49) into account.

Traveltime formula

With equations (3.20) and (3.23) describing the slowness vector projections and the above described four experiments, all 14 first- and second-order traveltime coefficients of equation (2.30) are expressed in terms of angles specifying the propagation direction of wavefronts at the measurement surface and matrices specifying the curvatures of these wavefronts at the measurement surface. Substituting these kinematic wavefield attributes into equation (2.30) yields

$$
\begin{aligned}
t\left(\mathbf{r_S}, \mathbf{r_G}\right) \approx t_0 - \mathbf{p_{S_0}} \cdot \mathbf{r_S} + \mathbf{p_{G_0}} \cdot \mathbf{r_G} \\
- \frac{1}{2v_S} \mathbf{r_S} \cdot \mathbf{R_S} \tilde{\mathbf{K}}_S^{CR} \mathbf{R}_S^T \mathbf{r_S} - \mathbf{r_S} \cdot \mathbf{E}\, \mathbf{r_G} + \frac{1}{2v_G} \mathbf{r_G} \cdot \mathbf{R_G} \tilde{\mathbf{K}}_G^{CS} \mathbf{R}_G^T \mathbf{r_G}.
\end{aligned}
\tag{3.51}
$$

The elements of $\mathbf{E}$ expressed in terms of wavefront curvatures are bulky equations. They are (for the sake of completeness) given in Appendix C.

3.5.2 Zero-offset case

In Section 2.5 it has been shown that the second traveltime derivatives in the ZO case can be described by two symmetric matrices. These matrices are given in the form of products of submatrices of the one-way surface-to-surface propagator matrix $\mathbf{T_0}$ (see equation (2.37)). Therefore, wavefront curvature matrices of two different waves detected at the measurement surface in the vicinity of the coinciding starting and end point of the central ray are required to relate the second traveltime derivatives to wavefront curvatures. The starting/end point of the central ray is referred to as Q_0 in the following.

The two linear coefficients in the ZO case are represented by the first two components of the slowness vector $\hat{\mathbf{p}}_0$ of the central ray at Q_0, where $\hat{\mathbf{p}}_0$ is given by

$$
\hat{\mathbf{p}}_0 = \frac{1}{v_0} \hat{\mathbf{R}}\, \tilde{\mathbf{e}}_z = \frac{1}{v_0} \begin{pmatrix} \cos\alpha_0 \sin\beta_0 \\ \sin\alpha_0 \sin\beta_0 \\ \cos\beta_0 \end{pmatrix}.
\tag{3.52}
$$

In equation (3.52), v_0 denotes the wave propagation velocity in the vicinity of Q_0. The matrix $\hat{\mathbf{R}}$ describes again the transformation from the global to the ray-centred coordinate system. It results from the matrix product $\hat{\mathbf{R}}_z\left(\alpha_0\right) \hat{\mathbf{R}}_y\left(\beta_0\right)$. The explanation of the matrix $\hat{\mathbf{R}}$ and the angles α_0 and β_0 can be found in Section 3.2. Note that $\hat{\mathbf{p}}_0$ as well as $\hat{\mathbf{R}}$ are associated with the direction of the central ray at its end point (and not at the starting point) which influences the range of β_0.

For the ZO case equations (3.21) and (3.22) can be written as

$$
\mathbf{p_G} \approx \mathbf{p_0} + \frac{1}{v_0} \mathbf{R}\, \tilde{\mathbf{K}}_G \mathbf{R}^T \mathbf{r_G}
\tag{3.53}
$$

and

$$
\mathbf{p_S} \approx -\mathbf{p_0} + \frac{1}{v_0} \mathbf{R}\, \tilde{\mathbf{K}}_S \mathbf{R}^T \mathbf{r_S},
\tag{3.54}
$$

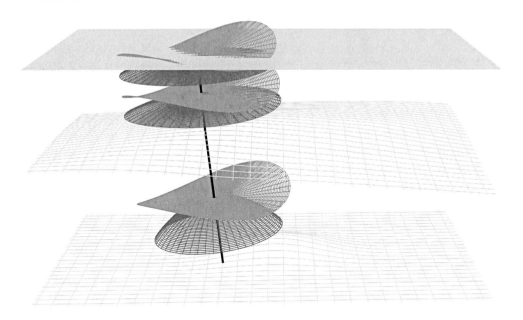

Figure 3.4: Model consisting of two iso-velocity layers separated by curved interfaces. The wavefronts of the NIP (in dark grey) and normal (in medium grey) wave propagating along the central ray (in black) are depicted at different points in time.

where $\mathbf{p_0}$ is the vector of the first two components of $\hat{\mathbf{p}}_0$ and $\mathbf{R}$ is the upper left 2×2 matrix of $\hat{\mathbf{R}}$. Note that equation (2.31) has already been taken into account in equation (3.54). The two latter equations will in the following be used to relate the second-order traveltime coefficients to wavefront curvatures.

Common-midpoint experiment (normal-incidence point wave experiment)

Hubral (1983) introduced two hypothetical experiments which are particularly elegant to relate second traveltime derivatives to wavefront curvatures. The first one is the so-called NIP wave experiment. This can be performed by placing a point source at the NIP of the central ray. The wave initiated in this way propagates along the central ray to the measurement surface. At Q_0 the propagation direction of its wavefront is given by $\hat{\mathbf{p}}_0$ and its wavefront curvatures are described by the matrix $\tilde{\mathbf{K}}_{\mathrm{NIP}}$. In Figure 3.4 the NIP wave experiment is depicted for a simple subsurface model consisting of two layers with constant wave propagation velocities. The dark grey surfaces are the NIP wavefronts which propagate along a normal ray to the measurement surface shown at different points in time.

The NIP wave experiment can also be formulated as a two-way experiment resulting in the identical wavefront as in the one-way experiment introduced above. Suppose a wavefront starts at Q_0

with curvatures defined by the matrix $-\tilde{\mathbf{K}}_{\mathbf{NIP}}$. After propagation along the cental ray it focuses at the NIP, reflects at this point and returns to Q_0, where the propagation direction is given by $\hat{\mathbf{p}}_0$ and the curvature matrix of the wavefront is described by $\tilde{\mathbf{K}}_{\mathbf{NIP}}$. In Appendix D it is shown by the so-called NIP wave theorem that the NIP wave experiment is—within the paraxial approximation—equivalent to the CMP experiment introduced in the previous section for the special case of a normal central ray. This means that

$$\tilde{\mathbf{K}}_{\mathbf{S}}^{\mathbf{CMP}} = -\tilde{\mathbf{K}}_{\mathbf{NIP}} \quad \text{and} \quad \tilde{\mathbf{K}}_{\mathbf{G}}^{\mathbf{CMP}} = \tilde{\mathbf{K}}_{\mathbf{NIP}} \,. \tag{3.55}$$

All rays involved in the CMP experiment for the ZO case are subject to the condition

$$\mathbf{r_G} = -\mathbf{r_S} = \mathbf{h} = \begin{pmatrix} h_x \\ h_y \end{pmatrix} \,. \tag{3.56}$$

Taking equation (2.38) into account, this leads to

$$\Delta \mathbf{m} = \begin{pmatrix} 0 \\ 0 \end{pmatrix} \,. \tag{3.57}$$

Substituting equation (3.57) into equation (2.37) gives rise to the CMP traveltime formula

$$t_{\mathrm{CMP}}(\mathbf{h}) \approx t_0 + \mathbf{h} \cdot \mathbf{B_0^{-1}} \mathbf{A_0} \mathbf{h} \,, \tag{3.58}$$

where the first derivatives of this formula with respect to h_x and h_y are approximated by

$$\frac{d t_{\mathrm{CMP}}}{d \mathbf{h}} \approx 2 \mathbf{B_0^{-1}} \mathbf{A_0} \mathbf{h} \,. \tag{3.59}$$

Analogous to the general case, $\frac{d t_{\mathrm{CMP}}}{d \mathbf{h}}$ can be written as a sum of $\mathbf{p_S^{CMP}}$ and $\mathbf{p_G^{CMP}}$ (see equation (3.37)). Inserting equations (3.55) into equations (3.53) and (3.54), one can therefore formulate an approximation for $\frac{d t_{\mathrm{CMP}}}{d \mathbf{h}}$ as

$$\frac{d t_{\mathrm{CMP}}}{d \mathbf{h}} = \mathbf{p_S^{CMP}} + \mathbf{p_G^{CMP}} \approx \frac{2}{v_0} \mathbf{R} \, \tilde{\mathbf{K}}_{\mathbf{NIP}} \mathbf{R^T} \mathbf{h} \,. \tag{3.60}$$

Comparing the second-order coefficients of equations (3.59) and (3.60) yields

$$\mathbf{B_0^{-1}} \mathbf{A_0} = \frac{1}{v_0} \mathbf{R} \, \tilde{\mathbf{K}}_{\mathbf{NIP}} \mathbf{R^T} \,. \tag{3.61}$$

Consequently, the second-order coefficients with respect to h_x and h_y of the traveltime formula (2.37) are related to the curvatures of the NIP wave.

Zero-offset experiment (normal wave experiment)

The second hypothetical experiment introduced by Hubral (1983) is the normal wave experiment which is also discussed in detail in Hubral (1984). This experiment is equivalent to what is known as the exploding reflector experiment described in Loewenthal et al. (1976). It can be performed by densely covering the reflector with point sources in the vicinity of the NIP. If all those point sources explode at the same time, a wave is generated which has the same shape as the reflector surface in the vicinity of the NIP. Therefore, the corresponding rays are normal to the reflector and the generated wave is called normal wave. After propagation along the central ray, the normal wave emerges at Q_0, where the propagation direction is given by $\hat{\mathbf{p}}_0$, and the wavefront curvatures are defined by $\tilde{\mathbf{K}}_N$. An example for a normal wave is shown in Figure 3.4, where the propagating wavefront of the normal wave, depicted in medium grey, is shown at different points in time.

Analogous to the NIP wave experiment, the normal wave experiment can also be formulated as a two-way experiment yielding the same wavefront at Q_0 as the one described in the experiment above. Suppose the curvatures of a wavefront starting at Q_0 are given by $-\tilde{\mathbf{K}}_N$. Propagating the wavefront along the central ray down to the reflector the wavefront curvature matrix at the NIP would be identical to the curvature matrix of the reflector surface. After reflection the wavefront returns to the measurement surface, where the wavefront curvature matrix at Q_0 is given by $\tilde{\mathbf{K}}_N$. As there are only normal rays, i. e. ZO rays, involved in the normal wave experiment, it is justified to refer to the normal wave experiment as ZO experiment. The surface which is orthogonal to the ZO rays can locally be described by a surface which possesses the same curvatures as the wavefront of the normal wave. This means that

$$\tilde{\mathbf{K}}_S^{ZO} = -\tilde{\mathbf{K}}_N \quad \text{and} \quad \tilde{\mathbf{K}}_G^{ZO} = \tilde{\mathbf{K}}_N, \tag{3.62}$$

where the matrices $\tilde{\mathbf{K}}_S^{ZO}$ and $\tilde{\mathbf{K}}_G^{ZO}$ represent the local curvatures of the surface orthogonal to the ZO rays at the starting points and end points, respectively.

The ZO rays have to fulfil the condition

$$\mathbf{h} = \begin{pmatrix} 0 \\ 0 \end{pmatrix}, \tag{3.63}$$

which, according to equations (2.38) and (2.39), is equivalent to

$$\mathbf{r}_S = \mathbf{r}_G = \Delta\mathbf{m} = \begin{pmatrix} \Delta m_x \\ \Delta m_y \end{pmatrix}. \tag{3.64}$$

Inserting equation (3.63) into equation (2.37) yields the ZO traveltime

$$t_{ZO}(\Delta\mathbf{m}) \approx t_0 + 2\mathbf{p}_0 \cdot \Delta\mathbf{m} + \Delta\mathbf{m} \cdot \mathbf{D}_0^{-1}\mathbf{C}_0\Delta\mathbf{m}. \tag{3.65}$$

Calculating the derivatives with respect to Δm_x and Δm_y gives rise to the approximation of the first traveltime derivatives given by

$$\frac{dt_{ZO}}{d(\Delta\mathbf{m})} \approx 2\mathbf{p}_0 + 2\mathbf{D}_0^{-1}\mathbf{C}_0\Delta\mathbf{m}. \tag{3.66}$$

Using equation (3.64), the first traveltime derivatives can alternatively be written as

$$\frac{dt_{ZO}}{d(\Delta \mathbf{m})} \approx \begin{pmatrix} \frac{\partial t_{ZO}}{\partial x_S} \frac{dx_S}{d(\Delta m_x)} + \frac{\partial t_{ZO}}{\partial x_G} \frac{dx_G}{d(\Delta m_x)} \\ \frac{\partial t_{ZO}}{\partial y_S} \frac{dy_S}{d(\Delta m_y)} + \frac{\partial t_{ZO}}{\partial y_G} \frac{dy_G}{d(\Delta m_y)} \end{pmatrix} = \begin{pmatrix} \frac{\partial t_{ZO}}{\partial x_S} + \frac{\partial t_{ZO}}{\partial x_G} \\ \frac{\partial t_{ZO}}{\partial y_S} + \frac{\partial t_{ZO}}{\partial y_G} \end{pmatrix} = -\mathbf{p}_S^{ZO} + \mathbf{p}_G^{ZO}. \qquad (3.67)$$

Substituting $\mathbf{p}_S^{ZO}$ and $\mathbf{p}_G^{ZO}$ in equation (3.67) by their approximations given by equations (3.53) and (3.54) and taking equations (3.62) into account leads to

$$\frac{dt_{ZO}}{d(\Delta \mathbf{m})} = 2\mathbf{p_0} + 2\frac{1}{v_0}\mathbf{R}\,\tilde{\mathbf{K}}_\mathbf{N}\,\mathbf{R}^T\Delta \mathbf{m}. \qquad (3.68)$$

Finally, comparing the second-order coefficients of equations (3.66) and (3.68) reveals that

$$\mathbf{D_0^{-1}C_0} = \frac{1}{v_0}\mathbf{R}\,\tilde{\mathbf{K}}_\mathbf{N}\,\mathbf{R}^T. \qquad (3.69)$$

Hence, the second-order coefficients with respect to Δm_x and Δm_y of traveltime formula (2.37) are formulated in terms of wavefront curvatures.

Traveltime formula

With the equations (3.52), (3.61), and (3.69), all eight first- and second-order traveltime derivatives for the ZO case are related to kinematic wavefield attributes. Considering these attributes in formula (2.37), the traveltime approximation for paraxial rays in the vicinity of the normal central ray can be written as

$$t(\Delta \mathbf{m}, \mathbf{h}) \approx t_0 + 2\mathbf{p_0} \cdot \Delta \mathbf{m} + \frac{1}{v_0}\Delta \mathbf{m} \cdot \mathbf{R}\,\tilde{\mathbf{K}}_\mathbf{N}\,\mathbf{R}^T\Delta \mathbf{m} + \frac{1}{v_0}\mathbf{h} \cdot \mathbf{R}\,\tilde{\mathbf{K}}_\mathbf{NIP}\,\mathbf{R}^T\mathbf{h}. \qquad (3.70)$$

The formulation of equation (3.70) in terms of $\mathbf{r_S}$ and $\mathbf{r_G}$ can easily be obtained by using equations (2.38) and (2.39).

The attractiveness of equation (3.70) is constituted by the possibility to directly deduce information about a reflector in the subsurface. Suppose the traveltimes along rays associated with a reflector below a homogeneous overburden, which are in the vicinity of the central ray, are known for all kinds of different offset and midpoint combinations. Then, from $\mathbf{p_0}$, $\tilde{\mathbf{K}}_\mathbf{NIP}$, and $\tilde{\mathbf{K}}_\mathbf{N}$, the local dips of the reflector at the NIP, the depth of the NIP, and the reflector's curvatures at the NIP can be determined. Therefore, traveltime formula (3.70) can be seen as a second-order approximation of the kinematic reflection response of a part of the reflector surface (the *common reflection surface*). If the reflector is below an inhomogeneous overburden, equation (3.70) still describes an approximation of the kinematic reflection response. Yet, the interpretation of the kinematic wavefield attributes associated with reflections from this reflector is not as straightforward as for homogeneous media. This means that their relations to the reflector's properties become complex. However, also for inhomogeneous media the attributes can be used to construct an image of the subsurface. Their use in seismic applications as well as their determination from seismic data will be elaborated in the following chapters.

3.5.3 Validity range of traveltime formulas

Traveltime formulas (3.51) and (3.70) both represent parabolic approximations of the correct traveltime which are valid in the vicinity of the central ray. The validity range of these formulas depends, next to the desired accuracy, on the complexity of the subsurface. It is, therefore, impossible to formulate a quantitative validity criterion without any knowledge of the subsurface. The use of kinematic wavefield attributes allows at least a qualitative examination of the validity range: the traveltime formulas (3.51) and (3.70) are valid as long as the wavefronts of the different experiments introduced in the previous sections are well approximated by second-order surfaces. In case of complex subsurface structures such as, for instance, rugged reflectors, fault zones, or strong near-surface velocity variations, where the wavefronts obtain complicated shapes, the validity range is diminished compared to simpler media.

Höcht (2002) investigated the validity of the use of kinematic wavefield attributes for second-order traveltime approximations in the ZO case by numerical experiments. For a medium with constant velocity layers separated by curved interfaces, he modeled the wavefronts of the NIP and normal wave along a normal ray from the reflection point to the emergence location at the measurement plane. Inserting the attributes determined in this way into equation (3.70) and comparing the traveltimes with traveltimes obtained from kinematic ray tracing, it was confirmed (at least for this model) that the traveltime approximation using kinematic wavefield attributes is indeed a good approximation of the exact traveltime.

3.5.4 Hyperbolic traveltime formulas

The parabolic traveltime formula is one of the possible second-order representations for approximating the correct traveltime. In fact, other second-order approximations can be formulated using the first- and second-order coefficients of equations (3.51) and (3.70). Another representation is given by a hyperbolic type of formula. It is obtained by squaring the parabolic equations and retaining only the first- and second-order terms. In the ZO case this yields

$$t_{\text{hyp}}^2(\Delta\mathbf{m},\mathbf{h}) = \left(t_0 + 2\mathbf{p_0}\cdot\Delta\mathbf{m}\right)^2 + \frac{2t_0}{v_0}\Delta\mathbf{m}\cdot\mathbf{R}\,\tilde{\mathbf{K}}_{\mathbf{N}}\,\mathbf{R}^{\text{T}}\Delta\mathbf{m} + \frac{2t_0}{v_0}\mathbf{h}\cdot\mathbf{R}\,\tilde{\mathbf{K}}_{\text{NIP}}\,\mathbf{R}^{\text{T}}\mathbf{h}\,. \tag{3.71}$$

In the seismic literature, equation (3.71) is often simply referred to as hyperbolic traveltime (indicated by the subscript *hyp*), recognising that it is an approximation of the correct traveltime. The term hyperbolic originates from the form of formula (3.71). In certain configurations and parameter combinations, equation (3.71) represents a hyperboloid. Note that expanding equation (3.71) into a Taylor series up to the second order yields again the parabolic approximation (3.70).

For the case of a plane horizontal or dipping reflector below a homogeneous overburden, the hyperbolic formula (3.71) is exact, whereas even in this simple case the parabolic equation (3.70) is only an approximation of the true traveltime. Moreover, Ursin (1982) and Gjøystdal et al. (1984) showed numerical examples which suggest that the hyperbolic traveltime approximation (3.71) is in many cases more accurate than the parabolic approximation (3.70). However, it is not possible to conclude that the hyperbolic formula is in general more accurate than the parabolic one. In fact, the difference in accuracy between the two formulas depends on the model under investigation.

3.5.5 Relationship between the 2D and 3D case

Traveltime formulas in terms of kinematic wavefield attributes for 2D media have been extensively discussed in many publications (see, e. g. Tygel et al., 1997; Höcht et al., 1999; Zhang et al., 2001). In this subsection I shortly explain the relationship between these 2D and the 3D formulas derived earlier in this chapter. For the 2D case I assume experiments, where starting and end points of all rays are located along a straight line on the measurement surface. This straight line has the same direction as vector $\hat{e}_x$. Therefore, all y-coordinates of the vectors in formulas (3.51) and (3.70) are zero.

General case

Under the assumption that the involved rays do not bend outside a plane which is perpendicular to the measurement surface, the angles α_S and α_G in equation (3.51) are zero. Substituting these 2D conditions into equation (3.51) yields

$$
\begin{aligned}
t\left(x_S, x_G\right) \approx t_0 &- \frac{\sin\beta_S^{2D}}{v_S}x_S + \frac{\sin\beta_G^{2D}}{v_G}x_G \\
&- \frac{\cos^2\beta_S^{2D}}{2v_S}k_S^{CR}x_S^2 - \frac{\cos^2\beta_G^{2D}}{2v_G}\left(k_G^{CMP}-k_G^{CS}\right)x_Sx_G + \frac{\cos^2\beta_G^{2D}}{2v_G}k_G^{CS}x_G^2,
\end{aligned}
\tag{3.72}
$$

where the angles β_S^{2D} and β_G^{2D} are given by

$$
\beta_S^{2D} = \beta_S \quad \text{and} \quad \beta_G^{2D} = \beta_G
\tag{3.73}
$$

and the curvatures k_S^{CR}, k_G^{CS}, and k_S^{CMP} are defined by

$$
k_S^{CR} = \tilde{k}_{S,00}^{CR}, \quad k_G^{CS} = \tilde{k}_{G,00}^{CS}, \quad \text{and} \quad k_G^{CMP} - k_G^{CS} = \tilde{k}_{G,00}^{CMP} - \tilde{k}_{G,00}^{CS}.
\tag{3.74}
$$

In equation (3.72) the relationship

$$
\frac{\cos^2\beta_S}{v_S}\left(\tilde{k}_{S,00}^{CR} - \tilde{k}_{S,00}^{CMP}\right) = \frac{\cos^2\beta_G}{v_G}\left(\tilde{k}_{G,00}^{CMP} - \tilde{k}_{G,00}^{CS}\right)
\tag{3.75}
$$

has been taken into account. Equation (3.75) can be derived using equation (B.9) presented in Zhang et al. (2001).

In case the rays do bend outside the plane which is perpendicular to the measurement plane (i. e. in case of out-of-plane reflections), the 2D attributes are more complicated functions of the 3D attributes. The angles are then expressed by

$$
\beta_i^{2D} = \arcsin\left(\cos\alpha_i \sin\beta_i\right),
\tag{3.76}
$$

and for the curvatures it follows that

$$
k_i^{(ex)} = \frac{\cos^2\alpha_i\cos^2\beta_i\,\tilde{k}_{i,00}^{(ex)} - 2\cos\alpha_i\sin\alpha_i\cos\beta_i\,\tilde{k}_{i,01}^{(ex)} + \sin^2\alpha_i\,\tilde{k}_{i,11}^{(ex)}}{1 - \cos^2\alpha_i\sin^2\beta_i},
\tag{3.77}
$$

The subscript 'i' stands for S or G, whereas the superscript 'ex' represents CR, CS, or CMP. Yet, the five-parameter form of equation (3.72) remains the same. Note that, although the curvatures of 3D wavefronts enter into the traveltime approximation, it is impossible to determine the full 3D information of the considered wavefronts by a 2D experiment.

ZO case

Assuming that rays do not bend outside the plane which is perpendicular to the measurement plane, the angle α_0 in equation (3.70) is zero. Considering this condition and that the y-coordinates are zero in equation (3.70), it follows for the parabolic approximation in the ZO case

$$t\left(\Delta m_x, h_x\right) \approx t_0 + 2\,\frac{\sin \beta_0^{2D}}{v_0}\,\Delta m_x + \frac{\cos^2 \beta_0^{2D}}{v_0}\left(k_{\mathrm{NIP}}\Delta m_x^2 + k_{\mathrm{N}}\,h_x^2\right), \tag{3.78}$$

and for the hyperbolic approximation

$$t_{\mathrm{hyp}}^2\left(\Delta m_x, h_x\right) = \left(t_0 + 2\,\frac{\sin \beta_0^{2D}}{v_0}\,\Delta m_x\right)^2 + \frac{2t_0 \cos^2 \beta_0^{2D}}{v_0}\left(k_{\mathrm{N}}\Delta m_x^2 + k_{\mathrm{NIP}}\,h_x^2\right). \tag{3.79}$$

where the angle β_0^{2D} is defined by

$$\beta_0^{2D} = \beta_0, \tag{3.80}$$

and the curvatures k_{N} and k_{NIP} are expressed by

$$k_{\mathrm{N}} = \tilde{k}_{\mathrm{N},00} \quad \text{and} \quad k_{\mathrm{NIP}} = \tilde{k}_{\mathrm{NIP},00}. \tag{3.81}$$

In case of out-of-plane reflections, the angle and the curvatures β_0^{2D}, k_{N}, and k_{NIP} are given by equations (3.76) and (3.77) where now the subscript 'i' is equal to 0 and the superscript 'ex' may stand for N or NIP.

In Höcht (2002) the relationship between the 2D and 3D kinematic wavefield attributes for the ZO case is described in detail. There, the case is also considered when the line, along which the starting and end points of the rays are located, has an arbitrary orientation on the measurement surface (i. e. not necessarily has the direction of $\hat{\mathbf{e}}_x$). Moreover, Höcht (2002) points out that the traveltime approximations for the 2D case are valid even when the subsurface is three-dimensional. This fact will be considered again in Chapter 5.

Chapter 4

3D seismic data acquisition and processing

Kinematic wavefield attributes can be extracted from the measured wavefield whenever coherent events are observable in the recorded data. The number of determinable attributes then depends on the way the data are acquired. That is, before examining how to extract the kinematic attributes from reflection seismic data, it is necessary to know about the acquisition geometry of these data. This applies in particular to the determination of kinematic wavefield attributes from 3D seismic data. Therefore, I outline commonly employed 3D seismic acquisition geometries in the first part of this chapter.

Moreover, it is required to elaborate, where the kinematic wavefield attributes can be of interest to assist and facilitate seismic processing. For this purpose, I delineate 3D seismic imaging schemes used nowadays in the second part of this chapter and show how the application of kinematic wavefield attributes for the ZO case introduced in the previous chapter can be integrated into processing schemes or can be an alternative to these.

4.1 3D seismic acquisition geometries

Seismic data are given in the form of time series which are called seismic traces. Thereby, a trace represents the response of the elastic wavefield at a receiver (geophone in land data acquisition; hydrophone in marine data acquisition) due to a seismic source (for example, weight drop, vibrator, explosive in land data acquisition; air gun in marine data acquisition)[1]. In a digital recording, the trace is sampled at a fixed rate in time, the inverse of which is called the sampling interval. Each trace is characterised by the corresponding source and receiver position. In the course of this work, I assume a flat measurement surface, where the geometry of sources and receivers is defined relative to a global Cartesian coordinate system (see Section 2.4) with its origin lying in the measurement surface. Therefore, source and receiver locations are given by the two-component

---

[1] See Sheriff and Geldart (1995) for a more detailed description.

vectors $\mathbf{s} = (s_x, s_y)$ and $\mathbf{g} = (g_x, g_y)$, respectively. Identifying the source and receiver positions with starting and end points of rays introduced in Chapters 2 and 3, the relationships

$$\mathbf{r_S} = \mathbf{s} - \mathbf{s_0} \quad \text{and} \quad \mathbf{r_G} = \mathbf{g} - \mathbf{g_0} \tag{4.1}$$

can be established. The vectors $\mathbf{s_0}$ and $\mathbf{g_0}$, respectively, define the source and receiver positions of the so-called central trace, i. e., the trace on which a reflection associated with a central ray is recorded. In the ZO case $\mathbf{s_0}$ and $\mathbf{g_0}$ coincide. For many processing methods it is useful to express the geometry of sources and receivers by midpoint and half-offset coordinates (see Chapter 2), where the midpoint and half-offset vectors are defined by

$$\mathbf{m} = \frac{\mathbf{g} + \mathbf{s}}{2} \quad \text{and} \quad \mathbf{h} = \frac{\mathbf{g} - \mathbf{s}}{2}. \tag{4.2}$$

In a seismic survey, for each shot (i. e. the activation of a seismic source), an array of receivers is employed recording data in form of common-shot volumes (ensemble of traces which share the same shot point). By variation of the source and the receiver array positions, a five-dimensional data (hyper-)volume (s_x, s_y, g_x, g_y, t) (or (m_x, m_y, h_x, h_y, t) in midpoint/half-offset coordinates) of traces is acquired, where t denotes the recording time of a trace. Such a data volume is referred to as prestack data volume. Signals which are reflected from the same interface in depth can be recorded at different traces. If these signals are isolated spikes, connecting the positions of the spikes on the traces would form a continuous (hyper)-surface in the prestack data volume. This (hyper-)surface can be seen as the kinematic reflection response of the interface. Typically, the extent of the survey area is larger than the lateral extent of the subsurface portion under investigation. This measure ensures that also steeply dipping subsurface structures can be illuminated.

In order to obtain the best possible processing result, it is, in principle, desirable that the prestack data volume is fully sampled. This means that the whole measurement surface is densely covered with both, sources and receivers, so that each possible source/receiver combination is available. In this way, a high data redundancy would be gained. That is, reflector points are illuminated through a multitude of experiments associated with different shot/receiver coordinates. For the midpoint/offset geometry, a fully sampled dataset has the consequence that each 3D CMP volume (i. e. an ensemble of traces which share the same midpoint) has a high fold (the number of traces in the CMP volume) and traces with all kinds of h_x- and h_y-coordinates, i. e. a full azimuthal coverage. Only the data redundancy, in fact, renders many important seismic processing methods possible such as, for example, stacking to improve the signal-to-noise ratio or velocity analysis (Yilmaz, 2001a). Also the algorithms to determine kinematic wavefield attributes from the prestack data as well as the 3D CRS stack introduced in the next chapter make use of this data redundancy.

Due to the prohibitively high costs of a fully sampled dataset, various 3D survey geometries are devised which constitute some compromise with respect to trace coverage and acquisition costs. The requirements on a seismic survey design to ensure that the acquired data contain the desired spatial continuity and resolution (to avoid aliasing) in order to enable the mappabilty of target structures of the subsurface and in order to give the possibility of noise suppression are discussed in Vermeer (1998). The classes of 3D data acquisition geometries can be roughly divided into line and areal geometries. Line geometries are used in marine acquisition utilising vessels and streamers as well as in land data acquisitions, whereas areal geometries are mainly applied in land surveys (ocean bottom cable/seismometer acquisition is not considered here). In a marine data acquisition, a vessel follows parallel sailing lines towing several (more or less parallel) streamers, where each streamer incorporates hundreds of receivers located at regular intervals. Normally, the sailing lines are aligned with the direction, where the strongest lateral inhomogeneities of the subsurface are expected. In the following it is assumed that this direction coincides with the x-direction of the global coordinate system used to describe the location of the traces. The x-direction is called the inline direction, whereas the y-direction, perpendicular to it, is termed the crossline direction. In case of a single streamer survey, a 3D marine survey is basically the repetition of a 2D survey, i.e., a survey conducted along a straight line. This means that for each sailing line there is only one line of midpoints, where for each CMP volume there are only traces in one azimuthal direction, namely the inline direction. Modern marine acquisition technology allows that one vessel tows up to 12 streamers and uses several sources which are alternately activated. In this way, for one sailing line several midpoint lines are produced, where for each CMP volume at least a small azimuthal range is covered with traces.

Line geometries are often also used for land data acquisition, where shot lines are deployed in parallel, orthogonal, or in an angle of $\pm 45°$ (zigzag) to the receiver lines (Vermeer, 1998). The shot and receiver lines are usually densely covered with sources and receivers, respectively, leading to a high fold in one azimuthal direction in the CMP volume. Yet, as in the marine acquisition, for these line geometries there is only a small azimuthal range covered with traces in one CMP volume. A full azimuthal coverage is virtually only achievable by a more evolved areal acquisition geometry. Usually, for this purpose an area is densely covered with receivers, but the shots are located only on a coarse grid (or vice versa, Vermeer, 1998). In this way the fold of a CMP volume is not considerably increased, but a wider azimuth range is covered. A commonly applied areal acquisition scheme is the swath shooting (Yilmaz, 2001b) in which receiver cables are laid out parallel to the inline direction and shots are positioned in lines perpendicular to the receiver lines, i.e. in crossline direction.

For both, the line and areal acquisition geometries, the prestack data volume has virtually always irregularities in the acquisition geometry. These occur, for instance, due to trees, rivers, buildings, etc. for the land data acquisition or cable feathering in the marine data acquisition. The irregularities complicate the transformation from shot/receiver to midpoint/half-offset coordinates (as is necessary for many processing methods) because the traces do not share exactly the same midpoint coordinates. Therefore, CMP binning (Yilmaz, 2001b) is required where neighbouring midpoints are gathered to one CMP bin and the associated traces form the CMP volume. However, CMP binning does not guarantee that the CMP fold is the same for each bin. Low fold areas should be identified because these can cause problems in the processing of the 3D data.

4.2   3D seismic imaging schemes

After some preprocessing (Yilmaz, 2001a), seismic data are subject to a multitude of different processing schemes. Seismic imaging aims at mapping the data recorded on the measurement surface into an image of the subsurface structures. The process of transforming reflection events from the time domain to their true subsurface location is known as migration (see, e.g., Gray et al., 2001; Biondi, 2003, for a tutorial treatment). In the world of seismic exploration, there are various classifications for migration methods:

- According to the way the migration result is achieved, there is the division into integral methods (e.g. Kirchhoff methods) and wavefield-continuation methods (Biondi, 2003).

- Migration methods are classified according to the type of input data, namely, prestack or poststack input data.

- Finally, migration methods are categorised by the output section, i.e. time or depth output sections.

Three-dimensional imaging has been, until today, dominated by the Kirchhoff method (particularly when it is applied to prestack data). Kirchhoff migration is performed by summing amplitudes along diffraction traveltime surfaces (see, e.g. Schneider, 1978; Schleicher et al., 2001) associated with samples in the output section and assigning the summation result to the respective output sample. Thereby, the diffraction traveltime surfaces can be computed using ray tracing. Wavefield-continuation methods are performed by numerical propagation of the recorded wavefield and formation of the subsurface image applying an imaging condition (Biondi, 2003). Both methods can be performed on poststack or prestack data. Poststack migration is done using the stacked data, where the amount of data is dramatically decreased compared to the prestack data, as the dimensionality of the data volume is reduced by stacking from five to three. Therefore, prestack migration (which uses all the traces that were acquired for migration) is computationally far more expensive than poststack migration. The desired output section for interpretation of the subsurface structures is mostly a depth section. For simpler subsurface structures, the depth-migrated image is just a stretched version of the time-migrated image. Since the use of time migration is less expensive than depth migration, time migration may be preferred in some situations.

There are several ways to produce the input for the poststack migration. A commonly used stacking technique is the 3D normal-moveout/dip-moveout (NMO/DMO) correction (Yilmaz, 2001b) followed by a stack over amplitudes which have the same traveltime to produce a ZO volume. The NMO/DMO correction removes the traveltime moveout from the traces of a CMP volume, i.e. the difference between the traveltime of an event on a central trace and any other trace in a CMP volume. In addition to a simple NMO correction, this procedure takes the reflector dip into account. Another stacking technique is the CMP stack, where in the 3D case the data in a CMP volume are stacked along hyperbolic traveltime surfaces to construct a ZO volume. The CMP stack is discussed in detail in the subsequent chapter.

Common to all migration schemes is the need for a velocity model defined either in depth or time. In fact, the quality of the migration result strongly depends on the accuracy of the velocity model. Therefore, a lot of effort is put into the determination of the velocity model. For simple media, a popular method to estimate the velocity model is based on the analysis of offset-dependent moveouts of the hyperbolic traveltime surfaces used for the 3D CMP stack. In case of a horizontally layered medium, the moveout of a reflection event in the CMP volume is approximately defined by $1/v_{\text{rms}}^2$, where v_{rms} denotes the root-mean-square velocity. It represents a weighted average of the wave propagation velocity within each layer (interval velocity) above the considered reflector and is associated with the ZO traveltime of the respective hyperbolic traveltime surface. The root-mean-square velocity model is on the one hand utilised for time migration and on the other hand serves as input to a Dix-inversion (Dix, 1955) to compute the interval velocities. This interval velocity model is then used for depth migration.

For more complex media, more advanced velocity estimation approaches are necessary. A class of velocity analysis methods, which are suited for complex media, are tomographic approaches, such as reflection tomography (see, e. g. Stork and Clayton, 1991) or stereotomography (see, e. g. Billette and Lambaré, 1998; Billete et al., 2003). These approaches are based on the principle of minimising the misfit between modeled and observed parameters to construct the velocity model. In case of reflection tomography the misfit between traveltimes along rays in the velocity model and the observed traveltimes is minimised by updating the velocity model. In stereotomography additionally the dips of traveltime curves are used. For 3D data, however, the huge amount of picking to get access to the traveltimes of the reflection events renders these methods difficult.

Yet another class of velocity analysis methods, known as migration velocity analysis (see, e. g. Al-Yahya, 1989; Liu and Bleistein, 1995), is directly based on the application of migration. This approach uses the flatness of reflection images in common-image gathers after prestack depth migration to estimate the velocity model where the common-image gathers represent the multi-coverage data after prestack depth migration for each midpoint sorted with respect to the offset of the involved traces. If the correct velocity model is estimated, the reflector images associated with primary reflections are flat in the common-image gathers. If not, the residual moveouts in the common-image gathers are picked, the velocity model is updated on the basis of the residual move-outs, and the data are migrated again with the new velocity model. Therefore, migration velocity analysis is in general an iterative procedure, particularly in the case of a complex subsurface.

It is commonly stated that 3D prestack depth migration is the most expensive procedure but the best solution to image complicated subsurface structures (see, e. g. Ratcliff et al., 1994; Jones et al., 1998; Rietveld and Summers, 2002). This is because in such situations stacking along approximate traveltime surfaces is seen as inadequate to produce an accurate ZO volume which degrades the poststack migration result. However, the statement presupposes that the velocity model for migration was accurately determined and the stack, as input to the poststack, migration was obtained applying one of the stacking techniques mentioned above. In case of a low signal-to-noise ratio and an irregular sampling of the prestack data, the velocity model determination is a difficult task. As shown below, in this regard, the 3D CRS stack and its kinematic wavefield attributes can be of help with respect to, firstly, the improvement of the image quality of the ZO volume and, secondly, the velocity model building process.

The improvement of image quality of the ZO volume compared to the NMO/DMO/stack sequence and the CMP stack can be explained as follows: in contrast to the CMP stack where traces from one CMP volume are considered during the stack, the CRS stack is not restricted to one volume but can make use of traces in the prestack data with arbitrary source/receiver locations. This is accomplished by stacking amplitudes of the traces along a hyper-surface which is defined by equation (3.71) (the implementation of the CRS stack is explained in detail in Chapter 5). In this way, a lot more traces are involved in the CRS stack than in the CMP stack. This is the reason why the CRS stack is expected to yield a considerably increased signal-to-noise ratio compared to the CMP stack. As opposed to the 3D CMP stack, where the hyperbolic traveltime surface is defined by three parameters (see, e. g. Yilmaz, 2001b), the 3D CRS stacking formula (3.71) is expressed by eight kinematic wavefield attributes. Therefore, the 3D CRS stack involves the determination of eight parameters before stacking. This makes the 3D CRS stack computationally far more expensive than the 3D CMP stack. The search for the eight parameters is, however, as robust as a commonly applied stacking velocity analysis and also works, as it is shown in Chapter 7, on irregularly sampled data. Moreover, compared to the NMO/DMO/stack sequence, the CRS stack may give a more accurate ZO volume, because the CRS stacking formula (3.71) leads in many cases to a better fit to the actual reflection responses than the applied NMO/DMO/stack operators (Müller, 1999).

The determination of the kinematic wavefield attributes does not only serve to formulate the 3D CRS stacking formula. The attributes are also of use to construct velocity models. Höcht et al. (2003) describe an algorithm to compute layered velocity models by means of the kinematic wavefield attributes where the (homogeneous) layers are separated by curved interfaces. This approach can be seen as a generalisation of the Dix-inversion mentioned earlier. In Duveneck and Hubral (2002) and Duveneck (2004) for the 2D case and indicated in Duveneck (2003) for the 3D case, a tomographic approach which makes use of kinematic wavefield attributes is introduced to estimate a smooth velocity model. This model is ideal for ray tracing and is, therefore, well suited for Kirchhoff migration. The picking of kinematic wavefield attributes as input to the tomography is done in the stacked ZO volume which has a significantly higher signal-to-noise ratio than the prestack data. Moreover, only few picks are required. Note that for each pick an approximation of the traveltimes of reflection events in the prestack volume is immediately available by means of the kinematic wavefield attributes and, therefore, the redundancy of the prestack data is implicitly used. Thus, compared to the two tomographic approaches mentioned above (reflection tomography and stereotomography), for this approach picking is considerably simplified. For a principle discussion on advantages and limitations of the method I refer to Duveneck (2004).

On the basis of the kinematic wavefield attributes, Mann et al. (2003) and Hertweck (2004) proposed an imaging workflow as an alternative to established schemes: determine the kinematic wavefield attributes from the prestack data, perform the CRS stack, estimate the velocity model using the attributes, and finally migrate the ZO data constructed by the CRS stack into depth. A variant to this scheme is to use the velocity model computed by means of the kinematic wavefield attributes, for prestack depth migration. In case of a low signal-to-noise ratio, where events in the common-image gathers are hardly detectable and migration velocity analysis is difficult, both approaches may considerably improve the migration result. The two proposed procedures can in fact be transferred to the 3D case. Yet, the application to 3D data has so far not been shown.

It should not go unmentioned that various applications and quantities, which are of interest for seismic imaging, are derivable from the kinematic wavefield attributes. For instance, first and second spatial traveltime derivatives, related to the attributes (see Chapter 3), are of use for traveltime interpolation, e. g. to support the computation of traveltime tables needed for Kirchhoff-type migration schemes (Vanelle and Gajewski, 2002a). Moreover, an approximation of the projected first Fresnel zone for ZO can be derived by the attributes. This zone represents the image on the measurement surface of that part of the reflector which influences a considered reflection (Hubral et al., 1993). It is, therefore, applicable as aperture limits for migration (Schleicher et al., 1997) and stacking (Vieth, 2001; Mann, 2002). Another important quantity for which an approximation can be calculated using the wavefield attributes is the modulus of the geometrical spreading factor along ZO rays (Hubral, 1983). It describes the change in amplitude of a wave due to propagation. Thus, the geometrical spreading factor can be used as a weight function to compensate propagation effects in stacked and migrated sections which are then termed true-amplitude sections (Hubral, 1983). The calculation of true-amplitude weights for migration by means of first and second traveltime derivatives are discussed in Vanelle and Gajewski (2002b). Yet another interesting application is a time migration directly based on the wavefield attributes. This can be conducted without any additional effort, once the kinematic wavefield attributes are determined. The theory and applications of this time migration for the 2D data case are shown in Mann (2002). For the 3D case, the basic formulas for this type of time migration are given in Appendix E. The discussion of practical aspects as well as the application of the attribute-based time migration are tasks yet to be done.

Chapter 5

Implementation of the 3D common-reflection-surface stack method

All applications of the kinematic wavefield attributes (introduced in Chapter 3) discussed in the previous chapter rely on an accurate determination of these attributes. Therefore, the search of the kinematic wavefield attributes will be elaborated in the following. This is done in connection with the 3D CRS stack to construct a ZO volume from multi-coverage data.

The search of the attributes involves a non-linear global eight-parameter optimisation problem. A simultaneous eight-parameter optimisation is, with regard to the computing time, practically unsolvable. Thus, it is necessary to devise a search strategy that represents the best possible compromise between an accurate attribute determination and computing time. The search algorithms presented below are derived from the approaches pointed out in Jäger (1999), Müller (1999), and Mann (2002) and describe the extension of these approaches to the 3D case.

5.1 Basic approach

The aim of the 3D CRS stack is to provide an accurate ZO volume with a high signal-to-noise ratio from multi-coverage prestack data using kinematic wavefield attributes. The principle approach for this task can be summarised as follows:

- Take the traveltime t_0 of a ZO trace of any primary reflection, where the coinciding source/receiver position is defined by the midpoint vector $\mathbf{m_0}$. In the following I will refer to this reflection as the central reflection.

- Determine the traveltimes of primary reflections with any combination of midpoint and offset coordinates which are associated with the same reflector in depth as the central reflection, i.e. the kinematic reflection response of the reflector.

53

- Sum (stack) the amplitudes along the hyper-surface defined by the kinematic reflection response in the five-dimensional prestack data volume and assign the summation result to t_0.

- Repeat the procedure for other central reflections at different sample positions defined by $(t_0, \mathbf{m_0})$. In this way, the desired ZO volume is produced.

- As the number of contributing traces may vary for different ZO samples, the stacked amplitude should be normalised by the number of traces contributing to the stack. This measure makes the different stacked amplitudes comparable.

The difficulty of this approach is that neither the ZO traveltimes of the central reflections nor the kinematic reflection responses are known. Moreover, the functions describing the kinematic reflection responses may become complicated, depending on the complexity of the subsurface.

The problem of the unknown ZO traveltimes is solved by defining a volume of grid points (samples) which all describe possible ZO traveltimes at different midpoint locations. Then, for each of these samples the stack is performed, irrespective of the fact whether the ZO traveltime is associated with an existing reflection or not. For the latter case there is no physically meaningful kinematic reflection response. Stacking along the hyper-surface defined by such an unphysical reflection response would sum only uncorrelated signals and, therefore, lead to a negligible stacked amplitude. Thus, only where the ZO traveltime is related to an actual reflection event, a significant stack amplitude is expected, and this without knowing these ZO traveltimes beforehand. Note that any coherent event contained in the prestack data is imaged by this approach, i. e., also (unwanted) events which are associated with multiple reflections.

The problem of the unknown kinematic reflection responses is tackled approximating these by a second-order traveltime formula as presented in Subsection 3.5.2. For the 3D CRS stack the hyperbolic traveltime (3.71) is preferable to the parabolic approximation (3.70), as the hyperbolic traveltime has in many cases shown to be more accurate than the parabolic one. Numerical experiments supporting this statement are given, for instance, in Höcht (1998) and Müller (1999) for the 2D case and in Ursin (1982) and Gjøystdal et al. (1984) for the 3D case. For recollection, the hyperbolic traveltime is presented here again:

$$t_{\mathrm{hyp}}^2 (\Delta\mathbf{m}, \mathbf{h}) = (t_0 + 2\,\mathbf{p_0} \cdot \Delta\mathbf{m})^2 + \frac{2t_0}{v_0} \Delta\mathbf{m} \cdot \mathbf{R}\,\tilde{\mathbf{K}}_{\mathbf{N}}\,\mathbf{R}^{\mathrm{T}} \Delta\mathbf{m} + \frac{2t_0}{v_0} \mathbf{h} \cdot \mathbf{R}\,\tilde{\mathbf{K}}_{\mathbf{NIP}}\,\mathbf{R}^{\mathrm{T}}\mathbf{h} \,. \qquad (5.1)$$

Traveltime curves or surfaces along which amplitudes are summed are often referred to as stacking operators. Therefore, equation (5.1) is termed the 3D CRS stacking operator. The eight parameters of equation (5.1) are searched for by means of coherence analysis. This is performed by variation of the eight parameters and the evaluation of the fit of the respective traveltime approximation to the kinematic reflection response in the prestack data. For this purpose, only traces within the range (aperture) where the kinematic reflection response is expected to be well approximated by equation (5.1) are considered. Once the parameter set yielding the best fit is determined, the stack can be performed using the traces located within the aperture. Thus, the 3D CRS stack provides, in principle, a ZO volume, eight parameter volumes, as well as a coherence volume. How to go about finding the eight parameters, will be explained in detail in the next section.

The upper part of Figure 5.1 shows a ZO volume which has been constructed from multi-coverage data by means of the 3D CRS stack. The corresponding velocity model used for the data modelling is depicted in the lower part of Figure 5.1. For each sample of the ZO volume, the eight parameters describing the 3D CRS stacking operator (5.1) have been determined. Note that for 3D seismic data the number of samples may easily amount to several hundred million.

5.2 Global optimisation problem

The search for the best-fitting 3D CRS operator for a chosen ZO sample $(t_0, \mathbf{m_0})$ involves a non-linear global optimisation problem with eight parameters in a five-dimensional data volume. Thereby, the fit of test operators is evaluated by computing the coherence of the prestack data along these operators. Thus, the coherence criterion defines the object function to be optimised. For real seismic data, the coherence as a function of the eight parameters shows, in general, a complicated behaviour with a multitude of local maxima which makes the localisation of its global maximum a complicated task. The coherence measure utilised in the following to determine the optimum fit of operators is the semblance criterion (Neidell and Taner, 1971; Douze and Laster, 1979). It is widely used in the seismic world for the detection of coherent events in the data. The definition of the semblance S is

$$
S = \frac{\sum\limits_{j=-W/2}^{W/2} \left(\sum\limits_{i=1}^{M} f_{i,j+w(i)} \right)^2}{M \sum\limits_{j=-W/2}^{W/2} \sum\limits_{i=1}^{M} f_{i,j+w(i)}^2} , \tag{5.2}
$$

where M is the number of traces contributing to the coherence analysis. The value of $w(i)$ describes the sample of the discrete CRS traveltime on the trace indexed by i. A time window of width W is centred about the CRS operator, where the index j represents the sample position in the time window. The time window accounts for the limited frequency bandwidth of the data and adds stability to the process. All amplitudes $f_{i,j(i)}$ within W enter into the semblance analysis, where $f_{i,j(i)}$ is calculated by a linear interpolation between the amplitude values associated with the two time samples next to $w(i)$. The sums in the denominator in equation (5.2) describe the energy in the subset of the prestack data used for the semblance analysis while the sums in the nominator yield the energy of the stacks along CRS operators shifted within the time window. Thus, the semblance coefficient S gives the normalised ratio of output to input energy which may vary between 0 and 1, where the maximum $S = 1$ is obtained if all considered amplitudes $f_{i,j(i)}$ are identical, separately for each fixed value of index j. Note that the semblance is decreased when the amplitudes vary along the CRS operator even when the CRS operator exactly describes the kinematic reflection response.

Let us suppose that for each of the eight parameters there is a range of 100 test values (which is no unrealistic number). Then, for every individual ZO sample, 100^8 possible 3D CRS operator representations exist. If the ZO volume contains 10^8 samples, the testing of all of these operators would implicate that the semblance coefficient S is computed 10^{24} times. Depending on the data acquisition and the used aperture limits, it is possible that thousands of traces are involved in the

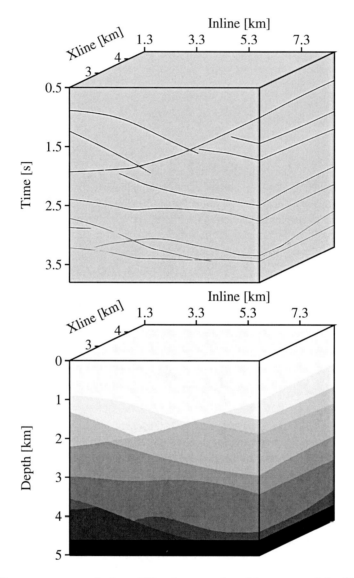

Figure 5.1: Upper part: example for a ZO volume produced by means of the 3D CRS stack. For each sample of the volume eight parameters have been determined. Note that the modelled prestack data contained only primary P-P reflections. Diffraction events were not considered during the data modelling. Lower part: depth model used for modelling which corresponds to the ZO volume depicted in the upper part, showing the distribution of P wave velocities which range between 2 (light grey) and 5 km/s (black).

coherence analysis of each 3D CRS operator. Therefore, such a simultaneous eight-parameter optimisation is, even on high-performance computers, completely impractical with regard to computing time. Thus, much faster search algorithms have to be devised.

For this purpose, the following measures can be taken:

- Sophisticated global optimisation methods can be utilised, such as simulated annealing (see, e. g. Kirkpatrick et al., 1983) which is a type of Monte-Carlo technique. These methods have the potential to reduce the number of tested parameter sets (and therefore the computing time) and—if correctly applied to the data at hand—still yield the global coherence maximum. However, their use in a simultaneous optimisation of eight parameters is far beyond any practicability. Even for global three-parameter optimisation, Mann (2002) concluded that the application of these methods is, from a practical point of view, "virtually unacceptable".

- The number of traces involved in the search may be reduced by decreasing the aperture. This, however, can only be done to a certain extent in order to keep the parameter determination stable, especially for noisy data. In any case, too large an aperture is not desirable, as the parameters to be determined should describe local properties of the kinematic reflection response (namely, first and second spatial derivatives of the traveltime).

- The parameter ranges may be constrained as far as possible. If available, a priori knowledge about the subsurface is of help in this respect.

- As already proposed in Müller (1999) and Mann (2002) for the 2D case, the global optimisation of all parameters can be decomposed into subsequent optimisations with less parameters using subsets of the full five-dimensional prestack data volume, such as the CS, CR, CMP, and CO configurations. Note that this measure is not at the expense of generality of the whole CRS approach and, as will be shown below, still yields accurate results.

In the following, the last three of the above mentioned points will be elaborated with regard to the way I actually implemented the 3D CRS stack, i. e. in the order of the processing steps. This is similar to the approach of Müller (1999), Jäger et al. (2001), and Mann (2002) proposed for the 2D case.

5.3 Processing steps

5.3.1 Common-midpoint configuration

The data configurations usually employed in seismic processing can be defined by a linear relationship between $\Delta\mathbf{m}$ and $\mathbf{h}$. Among these, the CMP configuration (defined by $\Delta\mathbf{m} = (0,0)^{\mathrm{T}}$) has one main advantage. It is the only configuration where equation (5.1) takes, instead of a five-parameter form, a three-parameter form, since the linear traveltime coefficients described by vector $\mathbf{p_0}$ cancel out[1]. Moreover, CMP binning is usually routinely applied to the prestack data so that traces belonging to a CMP configuration can easily be extracted from the prestack data. Therefore, it is particularly appealing to start the parameter search in the CMP configuration.

Inserting $\Delta \mathbf{m} = (0,0)^{\mathrm{T}}$ into equation (5.1) yields the CRS stacking operator for the CMP configuration

$$t_{\mathrm{CMP,hyp}}^2 (\mathbf{h}) = t_0^2 + \frac{2t_0}{v_0} \mathbf{h} \cdot \mathbf{R} \tilde{\mathbf{K}}_{\mathrm{NIP}} \mathbf{R}^{\mathrm{T}} \mathbf{h} . \tag{5.3}$$

This equation can also be formulated as

$$t_{\mathrm{CMP,hyp}}^2 (h_{\mathrm{x}}, h_{\mathrm{y}}) = t_0^2 + m_{00} h_{\mathrm{x}}^2 + 2 m_{01} h_{\mathrm{x}} h_{\mathrm{y}} + m_{11} h_{\mathrm{y}}^2 , \tag{5.4}$$

where the parameters m_{00}, m_{01}, and m_{11} can be explained by

$$\mathbf{M} = \begin{pmatrix} m_{00} & m_{01} \\ m_{01} & m_{11} \end{pmatrix} = \frac{2t_0}{v_0} \mathbf{R} \tilde{\mathbf{K}}_{\mathrm{NIP}} \mathbf{R}^{\mathrm{T}} . \tag{5.5}$$

From equation (5.4), the three-parameter form of the CRS operator (5.1) for the CMP configuration becomes obvious. In the 2D case, equation (5.4) reduces to a one-parameter hyperbola. In standard seismic processing, this parameter is determined by means of stacking velocity analysis (see, e. g. Yilmaz, 2001a). The same procedure can, in fact, also be applied to the 3D case. Let the half-offset vector $\mathbf{h}$ be expressed in terms of polar coordinates, i. e.

$$\mathbf{h} = r \begin{pmatrix} \cos \theta \\ \sin \theta \end{pmatrix} , \tag{5.6}$$

where r is the radial coordinate and θ the azimuth angle. Inserting equation (5.6) into equation (5.4), it follows that

$$t_{\mathrm{CMP,hyp}}^2 (r, \theta) = t_0^2 + \frac{4 r^2}{v_{\mathrm{stack}}^2 (\theta)} , \tag{5.7}$$

where $v_{\mathrm{stack}} (\theta)$ is the azimuth-dependent stacking velocity given by

$$\frac{1}{v_{\mathrm{stack}}^2 (\theta)} = \frac{t_0}{2 v_0} \begin{pmatrix} \cos \theta \\ \sin \theta \end{pmatrix} \cdot \mathbf{R} \tilde{\mathbf{K}}_{\mathrm{NIP}} \mathbf{R}^{\mathrm{T}} \begin{pmatrix} \cos \theta \\ \sin \theta \end{pmatrix} . \tag{5.8}$$

Equation (5.7) resembles the well-known CMP stack hyperbola of the 2D case. Thus, the three parameters in equation (5.4) can, in principle, be determined by standard stacking velocity analyses using CMP traces from three different azimuth directions. For example, the parameters m_{00}, m_{01}, and m_{11} relate to the three stacking velocities specified by $\theta = 0°$, $45°$, and $90°$ by

$$m_{00} = \frac{4}{v_{\mathrm{stack}}^2 (\theta = 0°)} , \tag{5.9a}$$

$$m_{11} = \frac{4}{v_{\mathrm{stack}}^2 (\theta = 90°)} , \tag{5.9b}$$

---

[1] The reason for this being the principle of reciprocity.

and

$$m_{01} = \frac{4}{v_{\text{stack}}^2 \left(\theta = 45°\right)} - 0.5 \left(m_{00} + m_{11}\right).$$ (5.9c)

The way I implemented the stacking velocity analysis is exemplarily shown for traces of a CMP volume from one fixed azimuth direction given by $\theta = 0°$ (see upper part of Figure 5.2). In order to define different stacking hyperbolas for testing, the range of traveltime moveouts (the traveltime difference to t_0) is discretised at the largest offset to be considered, i. e. at r_{ap} which specifies the limit of the aperture. The moveouts, along with the ZO traveltime, describe the stacking velocity of the respective hyperbola:

$$\frac{1}{v_{\text{stack}}^2} = \frac{i^2 \Delta t^2 - t_0^2}{4 r_{\text{ap}}^2} \quad \text{with} \quad i_{\min} \leq i \leq i_{\max},$$ (5.10a)

where

$$i_{\min} = \text{int} \left(\frac{1}{\Delta t} \sqrt{t_0^2 + \frac{4 r_{\text{ap}}^2}{v_{\text{stack,max}}^2} + 1} \right) \quad \text{and}$$ (5.10b)

$$i_{\max} = \text{int} \left(\frac{1}{\Delta t} \sqrt{t_0^2 + \frac{4 r_{\text{ap}}^2}{v_{\text{stack,min}}^2}} \right).$$ (5.10c)

Note that the term $(i\Delta t - t_0)$ expresses the traveltime moveout, where i is an integer. The function "int()" returns the integer of the value to which it is applied. The values of $v_{\text{stack,min}}$ and $v_{\text{stack,max}}$ are the lower and upper limits of the stacking velocity range tested for a ZO sample. These values are user-defined and may vary for different ZO samples, the way the stacking velocity varies for different values of t_0. Normally this means that both, $v_{\text{stack,min}}$ and $v_{\text{stack,max}}$, increase with increasing t_0. In case the average propagation velocity of a horizontally layered medium above the reflector is not smaller than the near-surface velocity v_0, v_0 is the lowest possible value for $v_{\text{stack,min}}$. For such a simple subsurface medium, events of multiple reflections which are associated with anomalously low stacking velocities can be attenuated by appropriately constraining the stacking velocity range (see, e. g. Mann, 2002).

By discretising the traveltime moveouts, the stacking velocity analysis can be connected with the temporal resolution of the data and, thus, with the highest achievable resolution. For instance, the increment of the traveltime moveout Δt can be chosen to be the sampling interval of the traces of the prestack data. Moreover, the test of less curved hyperbolas is not preferred as in the situation shown in the lower part of Figure 5.2. There, the same number of stacking hyperbolas within the same range is tested as in the upper part of Figure 5.2, yet, with a discrete increment of stacking velocities.

To accelerate the stacking velocity analysis, one may start with a large increment Δt and, therefore, perform a rough search for the stacking velocity. Afterwards, the search is refined by decreasing Δt, where only such stacking velocities are allowed which are close to the value found by the previous rough search. This procedure, however, is only applicable if the rough search already

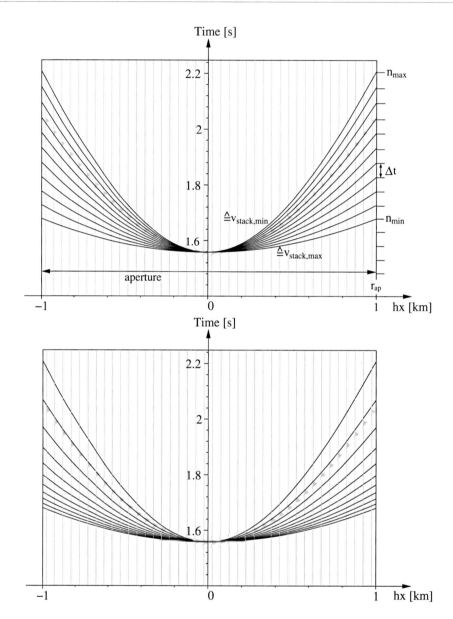

Figure 5.2: The upper part shows the stacking velocity analysis where the traveltime moveout at the largest considered offset is discretised to test different stacking hyperbolas. The values of $v_{stack,min}$ and $v_{stack,max}$ limit the range of tested hyperbolas. The lower part shows the stacking velocity analysis where test stacking velocities are incremented from $v_{stack,min}$ to $v_{stack,max}$ by a discrete stacking velocity value. Note that in the latter case, less curved hyperbolas are preferably tested.

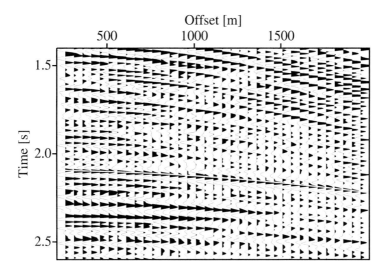

Figure 5.3: Part of a CMP gather of a marine dataset. The grey curve indicates the found stacking hyperbola associated with the found semblance maximum shown in Figure 5.4.

yields a stacking velocity close to the stacking velocity which corresponds to the global coherence maximum. For a CMP gather of a real seismic data set (see Figure 5.3), the semblance as a function of the stacking velocity for one ZO sample is depicted in Figure 5.4. This shows that the global coherence maximum is rather broad than sharp which endorses the practicability of the procedure. In fact, such a behaviour of coherence can be observed very often in stacking velocity analysis.

The vertical grey lines in the top graph of Figure 5.4 indicate the trial stacking velocities for which Δt is equal to the time sampling interval of the dataset. Note that the stacking velocity values are not equally sampled. The black vertical line represents the stacking velocity of the found coherence maximum. For the stacking velocity analysis in the middle graph of Figure 5.4, the increment of the traveltime moveout $\Delta t_{\mathrm{rough}}$ has been chosen seven times larger than for the upper graph and, therefore, the number of trial stacking velocities is seven times smaller. The dashed black line indicates the stacking velocity of the coherence maximum after the rough search. Obviously, this maximum is in the vicinity of the global coherence maximum, but not the optimum. Therefore, a refined search is performed which yields the stacking velocity given by the black solid line in the middle graph of Figure 5.4. This is attributed to the global coherence maximum. The way I perform the refinement of the search is illustrated in the bottom graph of Figure 5.4. In the first iteration, I vary the moveout of the hyperbola determined by the rough search at the largest considered offset by $\pm \Delta t_{\mathrm{rough}}/2$ and calculate the coherence for these two hyperbolas. The associated stacking velocities are labelled "1". For the second iteration, the moveout of the hyperbola which so far has given the highest coherence is varied by $\pm (\Delta t_{\mathrm{rough}}/2)^2$ and the coherence for these two hyperbolas (labelled "2") are computed. For the subsequent iterations the same procedure is car-

61

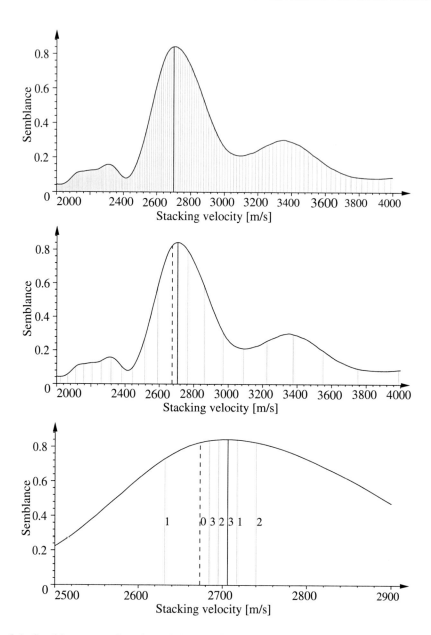

Figure 5.4: Semblance as a function of the stacking velocity. The increment of the traveltime moveout to test different stacking hyperbolas is 4 ms for the upper part and 28 ms for the middle part. The lower part shows the search refinement, where three iterations are performed. See text for more details.

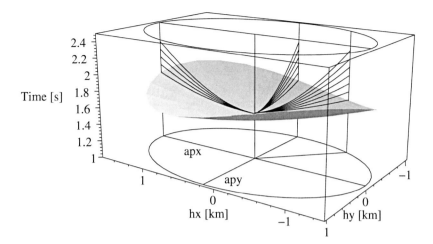

Figure 5.5: Stacking velocity analysis in three different azimuthal directions. In case of a three-parameter optimisation the fit of the operator (grey surface) is tested by simultaneous variation of stacking velocities in the three directions. The terms "apx" and "apy" indicate the aperture values in inline and crossline direction and define the major and minor axes of the aperture ellipse.

ried out, where the moveout variation is expressed by $\pm (\Delta t_{\text{rough}}/2)^n$, with n denoting the number of iterations. For the present example, three refinement iterations have been carried out which lead, together with the rough search, to a total number of 27 trial stacking velocities. This number is 5.3 times smaller than in the stacking velocity analysis shown in the upper part of Figure 5.4. The stacking hyperbola associated with the finally found stacking velocity is shown by the grey curve in Figure 5.3.

In case there is a sufficient number of traces in three different azimuthal directions to allow for a stacking velocity analysis, three one-parameter stacking velocity analyses can be performed in order to determine the three parameters of equation (5.4). However, this condition is not fulfilled for many multi-azimuthal data acquisition schemes. Therefore, a simultaneous optimisation of three parameters in the CMP configuration is necessary to ensure a stable determination of the CRS operator. The three-parameter search is depicted in Figure 5.5. The trial stacking velocities are tested for the sake of stability in three azimuthal directions specified by $\theta = 0°$, $45°$, and $90°$. For each of the directions, the range of stacking velocities can be limited. Due to the possible difference in complexity of the subsurface in different directions, this range may differ for each direction. The values of the stacking velocities are sampled in the same way as in the one-parameter stacking velocity analysis (see equations (5.10)). Yet, the fit of a surface (described by equation (5.4)) to the events in the prestack data has to be optimised. For each ZO sample, this is accomplished by testing all possible combinations of the trial stacking velocities for $v_{\text{stack}}(\theta = 0°)$, $v_{\text{stack}}(\theta = 45°)$, and $v_{\text{stack}}(\theta = 90°)$. With respect to computing time, it is reasonable to start the simultaneous three-parameter stacking velocity analysis with a rough search and refine the parameters afterwards, which can be done analogously to the one-parameter analysis described above.

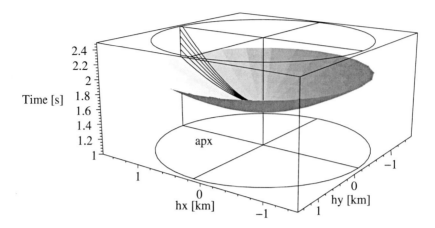

Figure 5.6: Stacking velocity analysis in one azimuthal direction using an operator which has a rotational symmetry with respect to the central trace. The value of "apx" defines the aperture limits.

In the case of a narrow-azimuth data acquisition as, for example, in marine measurements, a stable determination of all three parameters in equation (5.4) is impossible. In general, it is advisable to search for only one parameter. Figure 5.6 shows how this can be done. A stacking surface, described by one parameter, is varied and the best fit to the reflection events in the prestack data is evaluated by coherence analysis. Thereby, the surface is defined by

$$t^2_{\text{CMP,hyp}}(r) = t^2_0 + \frac{4\,r^2}{v^2_{\text{stack}}}, \tag{5.11}$$

where now the stacking velocity v_{stack} is azimuth-independent making the surface rotationally symmetric. The test of different stacking velocities is performed as explained before. The found stacking velocity is attributed to the inline direction. It would, in principle, be possible to gather traces of one azimuthal direction and try to find a best fitting hyperbola instead of a surface to determine v_{stack}. However, in the event of a streamer survey in the marine acquisition, the traces are, in most cases, not perfectly aligned in one azimuthal direction. The stacking surface also allows the use of traces which deviate from the predominant azimuthal direction. This stabilises the search for v_{stack}.

Theoretically, the use of equation (5.11) is strictly speaking only justified if propagation velocity variations occur with depth only, i. e. for the 1D case. If it is nevertheless applied to data acquired above a complex 3D medium, it leads to false stacking velocities and may, therefore, cause severe artifacts on the ZO traces later on. For this reason, I suggest for a multi-streamer survey, where a certain azimuth range is covered with traces, to start with a one-parameter search along the stacking surface (5.11). Afterwards, a three-parameter stacking velocity analysis can be performed by permitting, for each of the stacking velocities in the three azimuthal directions, small variations around the stacking velocity found by the one-parameter search (for example 20%). Thus, the three-dimensionality of the subsurface has been taken into account to a certain extent and, at the same time, the stability of the search is guaranteed. However, apart from the stacking velocity which corresponds to the inline direction, the stacking velocities can only give a trend and, if used for further processing, should be handled with care.

An important issue, which has not been addressed so far, is the choice of appropriate aperture limits. This is influenced by two aspects. On the one hand, local parameters (second traveltime derivatives at the ZO sample) are sought-after which suggests to choose the aperture as small as possible. On the other hand, from a numerical point of view, the parameter determination is more stable, the more traces are involved. The second point is especially important for noisy data. In principle, to obtain accurate parameters, the aperture limits should be chosen such that the kinematics of reflection events within the apertures more or less resemble the hyperbolic traveltime (5.4). As the reflection events can be completely different for different datasets, no general quantitative criterion for the choice of the aperture size can be formulated. In order to account for the complexity of reflection events in different azimuthal directions, the apertures may vary with azimuth. Therefore, I use an elliptical aperture in the $h_x h_y$-plane of a CMP configuration, where the minor and major axes of the ellipse are given by the apertures in inline and crossline direction (see Figure 5.5). Actually, there is no theoretical justification for this aperture shape. However, the ellipse has been tested and proven itself to be useful in practice. Moreover, as the complexity of reflection events may vary with the associated ZO traveltime, the apertures are also allowed to vary with t_0.

Whether the apertures have been chosen properly, can be checked by subtracting the traveltime moveouts from the reflection events, thus, producing moveout-corrected gathers. If the events are flat at ZO traveltime, the parameter determination was successful and, therefore, the apertures were appropriate. An example for CMP traces from a single azimuth is shown in the upper part of Figure 5.7, and a synthetic example for multi-azimuth data is depicted in the lower part Figure of Figure 5.7. For both cases, the left-hand side shows the traces of the prestack data before moveout correction within the aperture limit (time-varying for the upper example) and the right-hand side depicts the same traces after moveout correction.

The choice of the aperture limits does not only influences the accuracy of the determined stacking velocities, but also the number of tested values, due to the way I implemented the stacking velocity analysis (see Figure 5.8). In the case of constant stacking velocity ranges and constant moveout increments but varying apertures, choosing larger aperture limits results in more hyperbolas being tested.

5.3.2   Zero-offset configuration

After having determined three parameters of the 3D CRS operator (5.1), the remaining five parameters are searched in the ZO configuration. Inserting $\mathbf{h} = 0$ into equation (5.1) yields the CRS operator for the ZO configuration

$$t_{\text{ZO,hyp}}^2(\Delta\mathbf{m}) = \left(t_0 + 2\,\mathbf{p_0} \cdot \Delta\mathbf{m}\right)^2 + \frac{2t_0}{v_0}\Delta\mathbf{m} \cdot \mathbf{R}\,\tilde{\mathbf{K}}_{\mathbf{N}}\,\mathbf{R}^{\text{T}}\Delta\mathbf{m}. \tag{5.12}$$

65

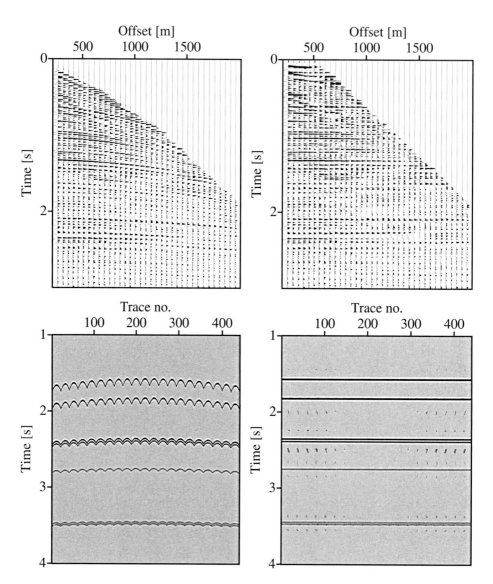

Figure 5.7: Upper part: CMP gather with traces from a single azimuth before (left) and after (right) moveout correction. Lower part: CMP gather with traces from multi-azimuth data before (left) and after (right) moveout correction.

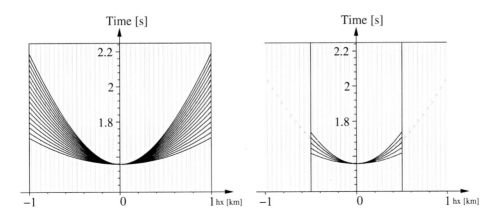

Figure 5.8: Effect of the aperture on the number of tested stacking velocity values: in case of a large aperture (left) far more values are tested than in the case of a small aperture (right).

Usually, ZO traces are not contained in the prestack data. For this reason, the ZO volume which is produced by the CMP stack is used. That is, the ZO volume is constructed from the prestack data by stacking traces along the CMP operator (5.4) within the apertures which have been used for the parameter search in the CMP configuration. Through this procedure, one takes advantage of the improvement of the signal-to-noise ratio achieved by the CMP stack, because the signal enhancement allows to separate the search of the five parameters in equation (5.12) as explained below.

In a first step, I assume that all coefficients related to second derivatives in equation (5.12) are zero. Expressed in terms of the kinematic wavefield attributes, this assumption means that the curvatures of the normal wave (see Chapter 3) are equal to zero. Therefore, the normal wave is approximated by a plane wave at the measurement surface. For this case, the traveltime approximation for the normal wave and, thus, the simplified CRS operator in the ZO configuration is given by

$$t_{\text{ZO,lin}}(\Delta \mathbf{m}) = t_0 + 2\mathbf{p_0} \cdot \Delta \mathbf{m} = t_0 + \mathbf{a} \cdot \Delta \mathbf{m} = t_0 + a_0 \Delta m_x + a_1 \Delta m_y, \qquad (5.13)$$

where the simple relations $2p_{0,x} = a_0$ and $2p_{0,y} = a_1$ for the components of $\mathbf{p_0}$ have been used.

Usually, the ZO traces obtained from the CMP stack lie more or less on a regular grid due to the CMP binning applied to the prestack data. Therefore, it is conceivable to separate the search for a_0 and a_1 into successive one-parameter searches using ZO traces from two azimuthal directions (for example, the inline and crossline direction). From the point of computing time, however, the simultaneous search for the coefficients a_0 and a_1 is also feasible. Moreover, more traces are involved in the surface-wise two-parameter search as depicted in Figure 5.9 than in the subsequently applied one-parameter searches. Hence, in case of noisy data, I prefer to use the simultaneous search.

In order to test several combinations of a_0 and a_1, the possible range of traveltime moveouts is discretised between the considered ZO sample t_0 and the traveltime at the aperture limits. Thereby,

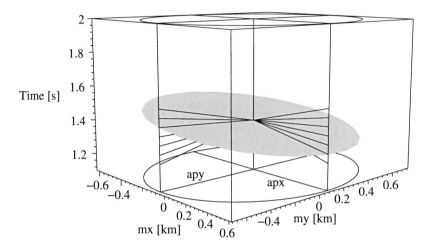

Figure 5.9: Determination of the linear traveltime coefficients in the ZO configuration: in case of a two-parameter optimisation, the fit of the operator (grey surface) is evaluated by simultaneous variation of different values for a_0 and a_1. The terms "apx" and "apy" indicate the aperture values in inline and crossline direction and define the major and minor axes of the aperture ellipse.

the aperture limits define the largest allowed difference between the central ZO trace and neighbouring ZO traces. The extremal traveltime moveouts have physical constraints. When looking at equations (3.52) and (5.13), it becomes obvious that $\pm 2/v_0$ are the extremal values for a_0 as well as a_1. This means that the slopes of the ZO events in any azimuthal direction may not exceed the slope of the wave traveling along the measurement surface with the propagation velocity v_0 that is reflected at an interface perpendicular to the measurement surface. Usually, these constraints are not strict enough, thus allowing too many implausible test values. Therefore, I further confine the extremal traveltime moveouts to $2 \sin \gamma / v_0$. The angle γ has a physical meaning, as it defines the steepest possible reflector dip for a reflector below a homogeneous medium. Hence, the tested values for a_0 and a_1 are described by

$$a = \frac{i \Delta t - t_0}{r_{ap}} \quad \text{with} \quad i_{min} \leq i \leq i_{max}, \tag{5.14a}$$

where

$$i_{min} = \text{int}\left(\frac{1}{\Delta t}\left[t_0 + \frac{2 r_{ap} \sin \gamma_{min}}{v_0}\right] + 1\right) \quad \text{and} \tag{5.14b}$$

$$i_{max} = \text{int}\left(\frac{1}{\Delta t}\left[t_0 + \frac{2 r_{ap} \sin \gamma_{max}}{v_0}\right]\right). \tag{5.14c}$$

The variable a stands for a_0 or a_1 and i is an integer. The value of r_{ap} defines, same as in equation (5.10), the aperture limit which is now the maximal midpoint dislocation in inline or crossline direction. Again, Δt is the increment of the traveltime moveout.

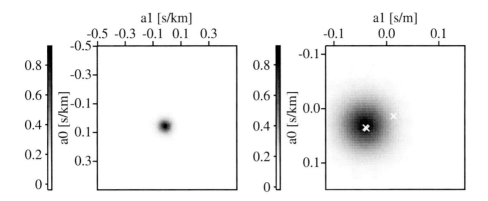

Figure 5.10: Semblance as a function of a_0 and a_1 for one ZO sample: for the shown example the semblance maximum is sharp. On the right-hand side, an enlargement of the region around the semblance maximum of the left panel. The white crosses indicate the locations of the semblance maxima after the rough search (uppermost white cross), after the search refinement (lowermost white cross), and the global semblance.

An example for the distribution of the semblance as a function of a_0 and a_1 for one ZO sample is depicted in Figure 5.10. This semblance panel has been computed for a simple synthetic dataset. However, it shows a characteristic which is also observable for real datasets: the coherence maximum is rather sharp which allows an accurate determination of a_0 and a_1 if the increment Δt is chosen properly. The three white crosses in the panel on the right-hand side of Figure 5.10 indicate the locations of the semblance values of the rough search (uppermost cross), after search refinement (lowest cross), and of the global coherence maximum (slightly above the value found by the search refinement). The value of Δt was ten times the traveltime sampling of the input data. Six iterations have been carried out for the search refinement which is implemented in the same way as described for the CMP configuration. For the example on hand, the processing values are selected such that the global maximum has been very closely met. With a smaller value for Δt during the rough search, a very slight improvement of the semblance is possible. In Figure 5.11 the determined linear traveltime approximations for the ZO sample at $(t_0 = 2.56\,\text{s}, m_{x,0}, m_{y,0} = 5000\,\text{m})$ of the inline and crossline section are displayed.

The accurate determination of a_0 and a_1 is also insofar important, as their quality influences the search of the second traveltime derivatives in equation (5.12). Reformulating equation (5.12) as

$$t_{\text{ZO,hyp}}^2 \left(\Delta m_x, \Delta m_y \right) = \left(t_0 + a_0 \Delta m_x + a_1 \Delta m_y \right)^2 + n_{00} \Delta m_x^2 + 2 n_{01} \Delta m_x \Delta m_y + n_{11} \Delta m_y^2, \quad (5.15)$$

where

$$\mathbf{N} = \begin{pmatrix} n_{00} & n_{01} \\ n_{01} & n_{11} \end{pmatrix} = \frac{2 t_0}{v_0} \mathbf{R} \, \tilde{\mathbf{K}}_{\mathbf{N}} \, \mathbf{R}^{\text{T}}, \quad (5.16)$$

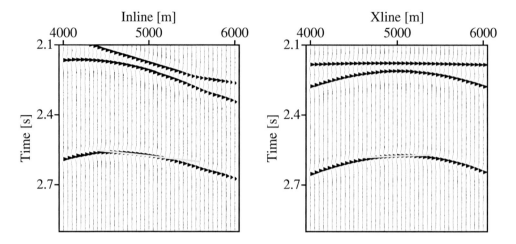

Figure 5.11: Found linear operators (indicated by grey lines) for one ZO sample in the inline (left) and crossline (right) section of a ZO volume associated with the found semblance maximum shown in Figure 5.10.

it becomes obvious that this search is a three-parameter optimisation problem which is similar to the three-parameter optimisation described for the CMP configuration. In the case of a simultaneous search, the fit of a surface to the reflection events is tested by all possible combinations of n_{00}, n_{01}, and n_{11}. The tested values are defined by the traveltime moveouts in three azimuthal directions at the aperture limits, where the moveouts are incremented by Δt from a minimal value to a maximal value (see Figure 5.12). Alternatively, the simultaneous three-parameter search can again be split into three successive searches, where now a shifted hyperbola is fitted to the reflection events along traces from three azimuthal directions. The shifted hyperbola as a function of the considered azimuth θ and the distance r (between the central ZO trace and the neighbouring ZO traces) is expressed by

$$t^2_{\text{ZO,hyp}}(r,\theta) = (t_0 + a(\theta)\, r)^2 + b(\theta)\, r^2 \tag{5.17}$$

where for $\theta = 0°$

$$\Delta m_x = r, \quad a_0 = a(\theta = 0°), \quad \text{and} \quad n_{00} = b(\theta = 0°), \tag{5.18}$$

for $\theta = 90°$

$$\Delta m_y = r, \quad a_1 = a(\theta = 90°), \quad \text{and} \quad n_{11} = b(\theta = 90°), \tag{5.19}$$

and for $\theta = 45°$

$$\Delta m_x = \Delta m_y = \frac{r}{\sqrt{2}}, \quad \frac{a_0 + a_1}{\sqrt{2}} = a(\theta = 45°), \quad \text{and} \quad n_{01} = b(\theta = 45°) - \frac{n_{00} + n_{11}}{2}. \tag{5.20}$$

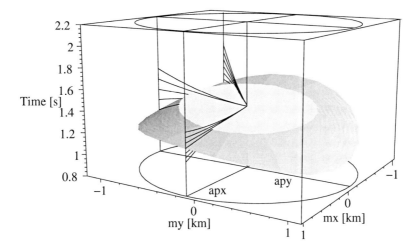

Figure 5.12: Determination of the second-order traveltime coefficients in the ZO configuration: in case of a three-parameter optimisation the fit of the operator (medium grey surface) is evaluated by simultaneous variation of different values for n_{00}, n_{10}, and n_{11}. The terms "apx" and "apy" indicate the aperture values in inline and crossline direction and define the major and minor axes of the aperture ellipse. The light grey surface represents the already determined linear operator.

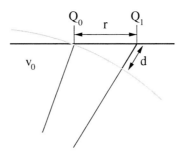

Figure 5.13: Wavefront of a normal wave shown in one azimuthal direction emerging at the measurement surface. If the near-surface velocity v_0 is constant, the traveltime for a wave along the distance d is shorter than for the distance r.

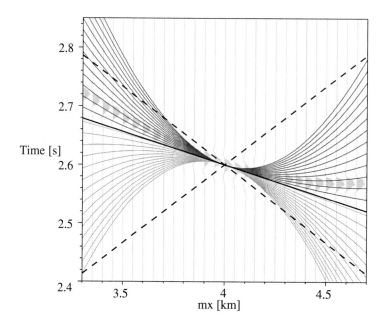

Figure 5.14: Search scheme for the second-order traveltime coefficients in the ZO configuration for one azimuthal direction. The dashed black lines indicate the extremal traveltime moveouts at the aperture limits. The solid black line represents the determined linear operator. The dark grey curves are defined by the ZO traveltime moveout at the aperture limit on the right hand side while the light grey curves are defined by the ZO traveltime moveout at the aperture limit on the left hand side.

The ranges of values for n_{00}, n_{01}, and n_{11} can be constrained due to the conclusions which can be drawn from Figure 5.13. There, a wavefront of the normal wave emerging at the measurement surface is depicted in an arbitrary azimuthal direction. Assuming a constant near-surface velocity v_0, the ZO traveltime moveout between the points Q_0 and Q_1 is given by $2d/v_0$ because the ZO ray starting and ending at Q_1 is, by the distance d, longer than the ray starting and ending at Q_0. For a wave propagating along the surface which is reflected at an interface that, in turn, is orthogonal to the measurement surface, the ZO traveltime moveout is expressed by $2r/v_0$. As the distance r is never smaller than the distance d, the traveltime moveout for any ZO reflection event is smaller than $\pm 2v_0$. Moreover, for the range of tested values, it has to be taken into account that the ZO traveltime surfaces are, in general, not symmetric to the location of the central ZO trace. Therefore, I calculate the considered traveltime moveouts partly at the largest positive aperture limits and partly at the largest negative aperture limits (see Figure 5.14). In this way, the highest achievable resolution can be obtained. Hence, the ranges of tested values for n_{00}, n_{01}, and n_{11} at the largest negative aperture limit are expressed by

$$n = \frac{i^2 \Delta t^2 - (t_0 - a\, r_{\mathrm{ap}})^2}{r_{\mathrm{ap}}^2} \quad \text{with} \quad i_{\min} \le i \le i_{\max}, \tag{5.21a}$$

where

$$i_{\min} = \mathrm{int}\left(\frac{1}{\Delta t}\left[t_0 + \frac{-2\,r_{\mathrm{ap}}}{v_0}\right]\right) \quad \text{if} \quad a < 0, \tag{5.21b}$$

$$i_{\min} = \mathrm{int}\left(\frac{1}{\Delta t}\left[t_0 + \frac{-a\,r_{\mathrm{ap}}}{v_0}\right]\right) \quad \text{if} \quad a \ge 0, \tag{5.21c}$$

$$i_{\max} = \mathrm{int}\left(\frac{1}{\Delta t}\left[t_0 + \frac{-a\,r_{\mathrm{ap}}}{v_0}\right] + 1\right) \quad \text{if} \quad a < 0, \quad \text{and} \tag{5.21d}$$

$$i_{\max} = \mathrm{int}\left(\frac{1}{\Delta t}\left[t_0 + \frac{2\,r_{\mathrm{ap}}}{v_0}\right] + 1\right) \quad \text{if} \quad a \ge 0. \tag{5.21e}$$

At the largest positive aperture limit the test values for n_{00}, n_{01}, and n_{11} are given by

$$n = \frac{i^2 \Delta t^2 - (t_0 + a\, r_{\mathrm{ap}})^2}{r_{\mathrm{ap}}^2} \quad \text{with} \quad i_{\min} \le i \le i_{\max}, \tag{5.22a}$$

where

$$i_{\min} = \mathrm{int}\left(\frac{1}{\Delta t}\left[t_0 + \frac{a\,r_{\mathrm{ap}}}{v_0}\right]\right) \quad \text{if} \quad a < 0, \tag{5.22b}$$

$$i_{\min} = \mathrm{int}\left(\frac{1}{\Delta t}\left[t_0 - \frac{2\,r_{\mathrm{ap}}}{v_0}\right]\right) \quad \text{if} \quad a \ge 0, \tag{5.22c}$$

$$i_{\max} = \mathrm{int}\left(\frac{1}{\Delta t}\left[t_0 + \frac{2\,r_{\mathrm{ap}}}{v_0}\right] + 1\right) \quad \text{if} \quad a < 0, \quad \text{and} \tag{5.22d}$$

$$i_{\max} = \mathrm{int}\left(\frac{1}{\Delta t}\left[t_0 + \frac{a\,r_{\mathrm{ap}}}{v_0}\right] + 1\right) \quad \text{if} \quad a \ge 0. \tag{5.22e}$$

In equations (5.21) and (5.22) the values of a and b for $\theta = 0°, 45°$, and $90°$ are explained in equations (5.18), (5.19), and (5.20). The variable r_{ap} denotes the magnitude of the aperture limit in the respective azimuthal direction. As can be seen in Figure 5.14, and as was already mentioned in Mann (2002) for the 2D case, with this procedure values are tested which are associated with operators that lie partly outside the theoretically defined region (dashed lines in Figure 5.14). However, this is not in contradiction to what was stated before. As the CRS operator is a second-order approximation, the operator will in any case fit to the reflection events only within a certain range. In other words, even those operators which slightly deviate from the theoretically predicted region might yield the highest coherence. In the worst case, as summarised by Mann (2002), the approach merely produces some "computational overhead".

Slices of a semblance volume (the value of n_{01} is fixed for each slice) for one ZO sample determined from a simultaneous optimisation of n_{00}, n_{01}, and n_{11} are shown in Figure 5.15. The coherence has a similar behaviour as for the CMP configuration. This means that the coherence maximum is rather broad compared to the sharp coherence maximum for the determination of a_0 and a_1 (see Figure 5.10), which is also observable for real data. In particular, the dependence of n_{01} on the semblance is weak, making this parameter difficult to determine. An explanation for this observation is that, for correctly determined values of a_0 and a_1, the operator (5.15) is tangent to the actual reflection event during the search for n_{00}, n_{01}, and n_{11}. Therefore, even for values of n_{00}, n_{01}, and n_{11} which slightly deviate from the correct values, the area of tangency yields a high coherence, leading to a broad semblance maximum.

As described for the CMP configuration, the three-parameter optimisation can be performed by a rough search followed by a refinement step. The white cross on the right hand panel in the middle of Figure 5.15 displays the location of the coherence maximum after the rough search, where Δt was ten times the traveltime sampling interval of the input data. After six refinement iterations, the location of the global semblance maximum was very well approximated, which confirms that the proposed search procedure was successful. The white cross on the left panel shows the location of the found semblance maximum after the refinement, whereas the location of the global semblance maximum is indicated by the white cross in the right panel on the top of Figure 5.15. Note that the values of n_{01} for the found and the global semblance maximum differ only very slightly. Figure 5.16 shows the found CRS operators as grey curves for the inline direction and the crossline direction of the ZO volume.

For the search apertures in the ZO configuration, the same considerations apply as for the CMP configuration: the apertures should be selected such that the CRS operator can be fit to the kinematics of the reflection events. Because the ZO volume reflects the complexity of the subsurface, the ZO apertures are in general smaller than for the search in the CMP configuration. As shown in Figures 5.9 and 5.12, the aperture limits in the $m_x m_y$-plane describe an ellipse, where the centre of the ellipse is located at the position of the central trace given by $\mathbf{m_0}$ and the minor and major axes are defined by the aperture limits in inline and crossline direction. In this way, the complexity of the medium in different azimuthal directions can be accounted for. The apertures for the search of a_0 and a_1 are normally considerably smaller than in case of the search for n_{00}, n_{01}, and n_{11}. The reason for this is that the range of the fit belonging to the linear operator (5.13) is smaller than the one for the hyperbolic operator (5.15). Analogously to the CMP configuration, the traveltime moveout corrections of the reflection events to the respective ZO traveltime of the central trace can be applied to check the choice of the aperture parameters.

5.3.3 Common-reflection-surface configuration

Using the eight parameters determined from the CMP and ZO configurations, the complete 3D CRS stack can now be performed in order to produce a ZO volume. This means that for each single sample of the ZO volume, traces from the five-dimensional prestack data volume are stacked along the 3D CRS operator (5.1) which can be formulated in terms of the eight determined parameters as (see equations (5.1), (5.5), (5.13), and (5.16))

$$t_{\text{hyp}}^2(\Delta\mathbf{m},\mathbf{h}) = \left(t_0 + \mathbf{a}\cdot\Delta\mathbf{m}\right)^2 + \Delta\mathbf{m}\cdot\mathbf{N}\Delta\mathbf{m} + \mathbf{h}\cdot\mathbf{M}\mathbf{h}. \tag{5.23}$$

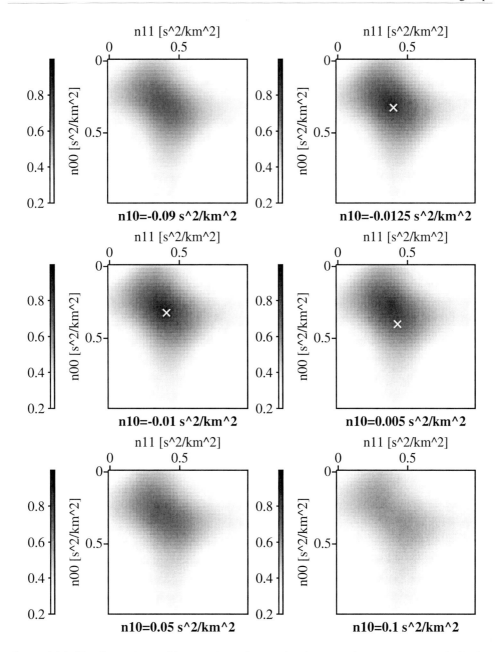

Figure 5.15: Six slices of a semblance volume from a simultaneous three-parameter optimisation of n_{00}, n_{10}, and n_{11} calculated for one ZO sample. The white cross in right panel on the top indicates the global semblance maximum. The white cross in the left panel in the middle indicates the semblance maximum after the search refinement; the white cross in the right panel in the middle shows the determined semblance maximum after the rough search.

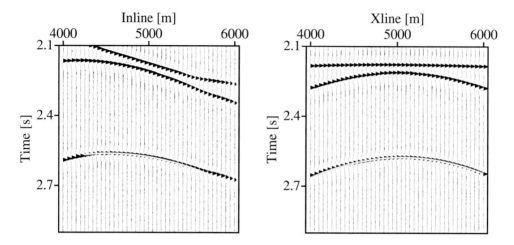

Figure 5.16: Found operators (indicated by grey lines) for one ZO sample in the inline and crossline section of a ZO volume associated with the found semblance maximum shown in Figure 5.15.

For the stack only those traces are considered which are located within the user-defined CRS aperture limits. These limits can be defined, for instance, by the apertures used for the searches in the CMP and ZO configuration. That is, the aperture limits of each of the CMP volumes which contribute to the CRS stack are limited by the search apertures for the CMP configuration, while only those CMP volumes are considered, where the corresponding ZO trace lies within the aperture limits used for the search in the ZO configuration. For complex data, it is advisable to decrease the aperture limits for a CMP configuration with increasing distance of the corresponding ZO trace to the central ZO trace.

In addition to performing the CRS stack, it is now possible to compute the kinematic wavefield attributes associated with each sample of the produced ZO volume from a_0, a_1, m_{00}, m_{01}, m_{11}, n_{00}, n_{01}, and n_{11} using equations (5.5), (5.13), and (5.16). For this purpose, the near-surface velocity v_0 has to be known. Note that for the parameter determination, v_0 is only used to constrain the search ranges. That is, even if a wrong value for v_0 is used, the 3D CRS stack is not affected, but only the kinematic wavefield attributes change their meaning. A synthetic data example, where the search strategy discussed in this section has been tested to extract the wavefield attributes and to construct a ZO volume, is shown in Chapter 6. The application of the proposed 3D CRS algorithms to a real dataset is presented in Chapter 7.

5.3.4 Further processing schemes and implementation aspects

The above proposed processing scheme is not the only way to determine the eight parameters of equation (5.1). In fact, there are many alternative processing schemes. One of these would be to start the parameter search in the CS configuration as this is the configuration in which the data are acquired. However, the CRS operator in the CS configuration has five parameters, where two of these are related to first traveltime derivatives. A simultaneous five-parameter optimisation is not applicable due to the required computing time. The separate determination of the two parameters associated with first traveltime derivatives in a first step is, in general, not applicable for real data as well, because the fit of linear operators to noisy prestack data is unstable. Actually, this is the reason why I choose not to start the parameter search with the CS configuration but with the CMP configuration. As already stated, this is the only configuration for which the CRS operator has no parameters associated with first traveltime derivatives.

Another alternative processing scheme was proposed in Höcht (2002). This scheme makes use of the fact that the 2D CRS operator (3.79) is also applicable when the medium varies in all azimuthal directions. Therefore, the eight parameters in equation (5.1) can be determined using three 2D CRS stacks which were produced from data acquired along three different lines on the measurement plane. This scheme would be fast and stable. However, it requires that for each ZO location there are three crossing acquisition lines with an appropriate azimuthal distribution. This is virtually never the case for real data with commonly used acquisition geometries.

For the refinement of the three-parameter searches in the CMP and ZO configurations, involved optimisation algorithms, such as the simulated annealing, can be used. In order to keep the computational costs at a reasonable level, Müller (2003) proposed to apply these algorithms as local optimisers using the attributes determined from the searches described above as initial values for the refinement.

An important implementation issue that has not been addressed so far is the parallelisation of the parameter search. Actually, for the everyday use of the 3D CRS stack, the parallelisation is mandatory. As a matter of fact, the parameter search is ideally suited for parallelisation as the parameters needed for one ZO sample can be calculated independently from any other sample. Several technical aspects for the parallelisation of the 3D CRS stack are discussed in Müller (2003).

Chapter 6

Test of the 3D common-reflection-surface stack implementation on synthetic data

Before applying the 3D CRS stack to real data, the search algorithms proposed in the previous chapter are tested. The synthetic data used for this purpose are rather idealised in comparison to real data. Nevertheless, the data are realistic enough to test whether the algorithms are applicable in practice.

6.1 Model and prestack data

The elastic model used for the implementation test consists of isotropic iso-velocity blocks bounded by curved interfaces. The P wave velocities range from 2000 to 5000 m/s. The S wave velocities are given dividing the P wave velocity by $\sqrt{3}$. The densities of the blocks (not displayed) range between 2000 and 4000 kg/m^3. Vertical slices of the velocity model are shown in Figures 6.1 (at fixed crossline positions) and 6.2 (at fixed inline positions). The figures show fault structures and a three-dimensional dome structure causing some stronger velocity changes in inline direction and only moderate variations in crossline direction.

For this model a multi-coverage dataset has been calculated by ray tracing, where only primary P waves were considered. Diffraction events have not been modelled. As source pulse a 30 Hz zero-phase Ricker wavelet was used. The traces are sampled at an interval of 4 ms. For the data modelling, changes of the amplitudes due to reflection and transmission at the interfaces were neglected. As acquisition geometry, a land data geometry was simulated, where shots and receivers were shifted along the measurement surface in a way to produce an optimal azimuthal coverage for the CMP volumes. The midpoint-offset geometry and recording parameters are listed in Table 6.1. Finally, band-limited random noise was added to the traces. In order to get an impression of the signal-to-noise ratio, some traces of a CS volume are plotted in Figure 6.3. Note that in some

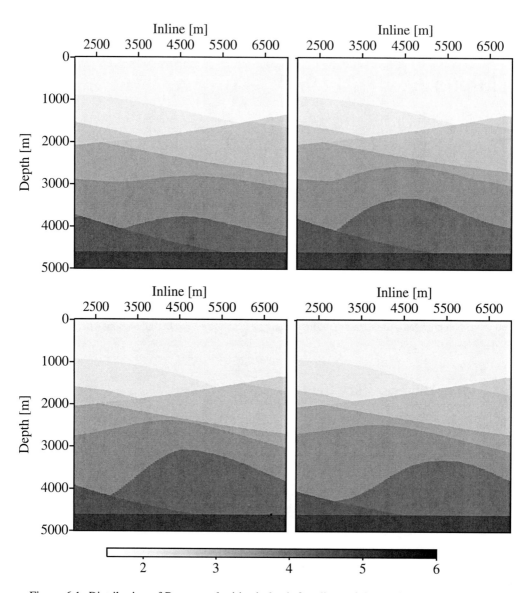

Figure 6.1: Distribution of P wave velocities in km/s for slices of the model for a fixed crossline coordinate at 3000 m (upper left figure), 4000 m (upper right figure), 5000 m (lower left figure), and 6000 m (lower right figure).

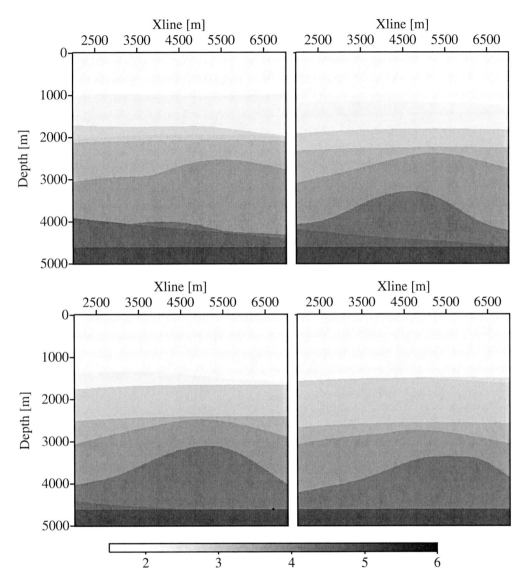

Figure 6.2: Distribution of P wave velocities in km/s for slices of the model for a fixed inline coordinate at 3000 m (upper left figure), 4000 m (upper right figure), 5000 m (lower left figure), and 6000 m (lower right figure).

Midpoint and offset geometry	
Number of CMP bins in inline direction	79
Number of CMP bins in crossline direction	79
CMP fold	25...441
Range of Cartesian CMP bin coordinate in inline direction	2550...6450 m
Range of Cartesian CMP bin coordinate in crossline direction	2550...6450 m
CMP bin interval in inline direction	50 m
CMP bin interval in crossline direction	50 m
Offset range	0...1000 m

Recording parameters	
Recording time	3.5 s
Sampling interval	4 ms
Peak frequency of zero-phase Ricker wavelet	30 Hz

Table 6.1: Acquisition parameters for the modelling of the multi-coverage dataset.

places there are gaps along the reflection events. These occur due to the way the modelling was performed. For each acquired shot volume, the model had to be confined to save computing time (which amounted to more than one week on six processors of an SGI Origin 3200) and, therefore, some small parts of reflection events, where the corresponding rays leave the model, are missing.

6.2 Processing parameters

Due to the full azimuthal coverage in most of the CMP volumes of the prestack data, a simultaneous three-parameter stacking velocity analysis was performed. The stacking velocity ranges were chosen to allow for the P wave interval velocities occuring in the model. For this, the stacking velocities cannot be larger than the P wave interval velocities. The apertures were selected such that all traces of the CMP volumes entered into the velocity analysis. The subsequent searches using the ZO volume produced by the CMP stack involved a simultaneous two-parameter optimisation of the linear traveltime coefficients of equation (5.13) and a simultaneous three-parameter optimisation of the quadratic traveltime coefficients of equation (5.15). The ZO apertures were selected after visual inspection of the CMP stack results and in such a way to account for the complexity of the ZO reflection events. Moveout increments of all optimisation steps were adjusted to enable a stable parameter estimation within a reasonable computing time. For the full parameter search, the total computing time was less than two days on a PC with a dual-processor system (each processor 1.4 MHz, 1 GB RAM). For the finally performed eight-parameter 3D CRS stack, the choice of apertures led to a range of 230 to 10000 prestack traces contributing to a single simulated ZO sample. The extensions of the ZO volume produced by means of the 3D CRS stack and utilised processing parameters are summarised in Table 6.2.

82

Output parameters	
Range of CMP coordinate in inline direction	2550…6450 m
Range of CMP coordinate in crossline direction	2550…6450 m
Simulated ZO traveltimes	0.9…3.5 s

General search parameters	
Near-surface velocity	2000 m/s
Total width of coherence time window	36 ms

Search parameters in CMP volume	
Number of search parameters	3 (simultaneous)
Moveout increment for rough search	24 ms
Number of search refinements	4
Stacking velocity range for all azimuths	1900…5000 m/s
Offset aperture in x-direction	1000 m
Offset aperture in y-direction	1000 m

Search parameters in ZO volume	
Number of search parameters	2 linear (simultaneous)
	3 quadratic (simultaneous)
Moveout increment for rough dip search	12 ms
Number of search refinements for dip search	6
Midpoint aperture in inline direction for dip search	560 m
Midpoint aperture in crossline direction for dip search	560 m
Moveout increment for rough curvature search	24 ms
Number of search refinements for curvature search	6
Midpoint aperture in inline direction for curvature search	800 m
Midpoint aperture in crossline direction for curvature search	800 m
Range of angle γ	$-30°…30°$

Stack parameters	
Offset aperture for each CMP bin in inline direction	600 m
Offset aperture for each CMP bin in crossline direction	600 m
Midpoint aperture in inline direction	500 m
Midpoint aperture in crossline direction	500 m

Table 6.2: Selected processing parameters.

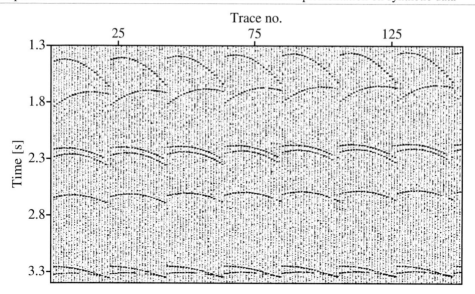

Figure 6.3: Traces of a common-shot volume extracted from the prestack data.

6.3   Processing results

During the processing, a multitude of stack and attribute volumes are constructed, all of which are useful for the evaluation of the chosen processing parameters and the interpretation of the imaging results. An example for a stack volume is shown in Figure 5.1. It would be possible to display such volumes also for every single attribute. For the presentation of the results, however, it is more convenient to display slices of the volumes. For this reason, I discuss the processing results on an inline section with a fixed crossline coordinate of 5000 m and on a crossline section with a fixed inline coordinate of 5000 m which are extracted from the different volumes. In the following I simply refer to these sections as inline and crossline sections.

The inline and crossline sections of the ZO volume constructed by means of the 3D CRS stack are shown in Figure 6.4. Compared to the traces of the prestack data presented in Figure 6.3, the signal-to-noise ratio is strongly enhanced and the reflection events are clearly visible. The gaps along the reflection events in the inline section are mainly due to deficiencies of the modelling of the prestack data, where, as mentioned above, the gaps are already present. The improvement of the signal-to-noise ratio and the reconstruction of the Ricker wavelet give a first indication that the CRS stacking operators have been well determined. This conclusion can be drawn because stacking 10000 traces along CRS operators which do not fit to the actual reflection events would most likely destroy the wavelets. Another indication that the correct CRS operators have been found is given by the semblance values shown in Figure 6.5. These values clearly reveal the detection of reflection events. For the example at hand, only those ZO samples are related to an actual event whose associated semblance is higher than 0.2. For most of these ZO samples, the

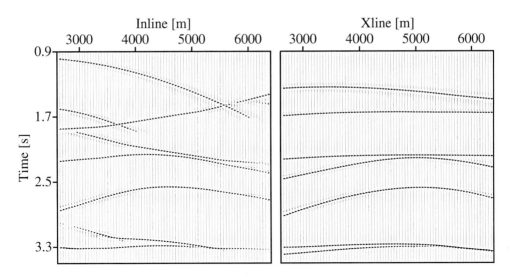

Figure 6.4: 3D CRS stack results: on the left-hand side an inline ZO section with a fixed crossline coordinate of 5000 m is shown, while on the right-hand side a crossline ZO section with a fixed inline coordinate of 5000 m is depicted. Both sections are extracted from the constructed ZO volume.

semblance is even higher than 0.7. This signifies the good fit of the CRS operators.

An example for an ensemble of traces which contribute to one CRS output trace is depicted in Figure 6.6. Such an ensemble of traces is often referred to as a 3D CRS gather. For clarity, only those segments of the traces are shown, where reflection events are present. For the respective subfigures, which have the same time window, the upper subfigure shows the time segment of the prestack traces before 3D CRS moveout correction, whereas the lower subfigure depicts the prestack traces after 3D CRS moveout correction. The wiggle plots on the right-hand side respresent the respective CRS stack results which can be obtained by stacking the uncorrected traces along the 3D CRS operator or by stacking the moveout corrected traces horizontally. In the moveout corrected gather the horizontal alignment of the events is almost perfect which is actually observable for every moveout corrected CRS gather. This horizontal alignment demonstrates that the parameters of the CRS operators associated with reflection events are well determined and that the apertures for the parameter searches and for stacking were selected properly. This means that the kinematic reflection responses of the individual interfaces can be well approximated by the hyperbolic traveltime formula (5.23) within the aperture limits. Finally, in the moveout corrected gather, there are no pulse stretch effects as occur, in general, after NMO correction (see Perroud and Tygel, 2004, for a detailed discussion). The explanation for this is given in Mann and Höcht (2003): due to semblance as the chosen coherence criterion, the 3D CRS operators are fit to isophase surfaces of the reflection events. Thus, the band-limited nature of seismic reflection data is implicitly considered leading to no pulse stretch.

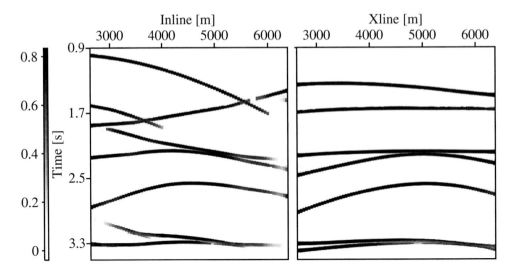

Figure 6.5: Coherence sections showing the semblance values associated with the ZO sections shown in Figure 6.4.

Yet another way to check if the parameter search was successful, is to compare the ZO traveltimes of the reflection events estimated with the 3D CRS stack with ZO traveltimes obtained by ray tracing, using the correct velocity model (see Figures 6.1 and 6.2). For this comparison, the amplitude maximum of each wavelet has to be picked. This maximum represents the traveltime along the associated ZO ray since a non-causal zero-phase Ricker wavelet was used for modelling. In Figure 6.7 traces from three different inline locations (which are indicated by the number on top of each subfigure) are presented. In each subfigure the left trace is extracted from the inline section of the ray-traced ZO volume, the middle trace is extracted from the ZO volume constructed by the 3D CRS stack, and the right trace is the difference between the two latter traces. In Figure 6.8 the same is depicted as in Figure 6.7 but for traces of the crossline section. Both figures show that the wavelets of the stacked traces resemble the modelled wavelets very well. The positions of amplitude maxima of the wavelets for the ray tracing and 3D CRS stack results are almost identical. The absolute value of the average difference between the positions of amplitude maxima is ≈ 4 ms, i.e., a difference is hardly detectable. This holds true for almost all traces of the ZO volume constructed by the 3D CRS stack. A resolution problem can be observed for the two wavelets around $t = 3.3 s$ in the subfigure on the right-hand side of Figure 6.7. For the trace obtained from ray tracing there are two peaks visible, whereas there is only one clear peak in the result of the CRS stack. This loss of resolution can be attributed to the time window which has been used during the coherence analysis to stabilise the parameter search (see Chapter 5). However, the overall impression is that the ZO traveltimes are well estimated by means of the 3D CRS stack for this data example.

With the knowledge of the near-surface velocity, all eight kinematic wavefield attributes for the

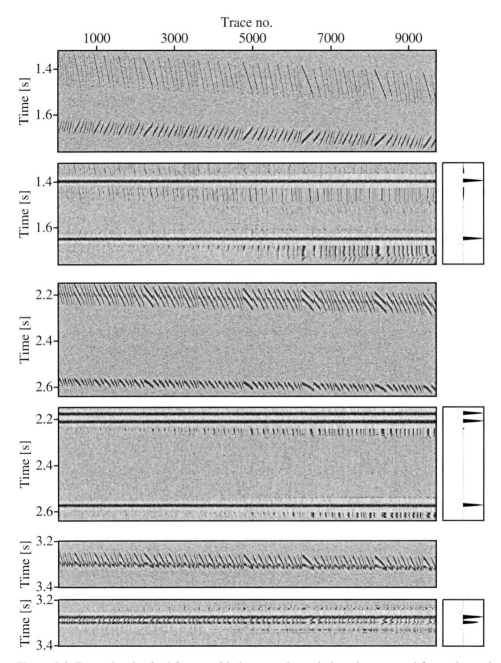

Figure 6.6: For each pair of subfigures with the same time window, the upper subfigure shows the time segment of the traces before moveout correction; the lower subfigure depicts the traces after moveout correction. The wiggle plots on the right-hand side respresent the CRS stack result.

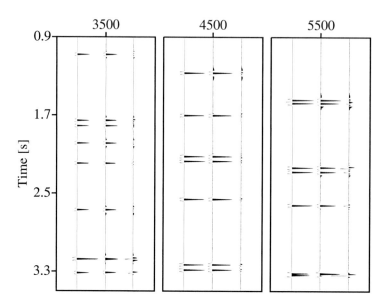

Figure 6.7: Traces from inline locations: in the subfigures (number at the top indicates the inline coordinate in m) the left trace is from the ray-traced ZO volume, the middle trace is from the ZO volume constructed by the 3D CRS stack, and right is the difference between the latter two.

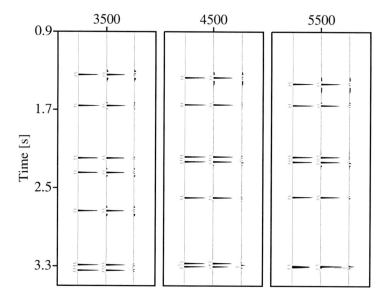

Figure 6.8: Traces from crossline locations: for each triplet the same applies as in Figures 6.7. The number at the top indicates the crossline coordinate in m.

ZO case can be calculated after the 3D CRS stack. For recollection, these attributes are the two angles which determine the propagation direction of the NIP and normal wave, as well as the three wavefront curvatures associated with the NIP and the three wavefront curvatures associated with the normal wave (see Chapter 3). Apart from the comparison between the ZO traveltimes obtained from ray tracing and ZO traveltimes estimated by means of the 3D CRS stack, in the case of real data, the stack sections, the coherence sections, the number of contributing traces, and the moveout corrected sections give the possibility to evaluate if the kinematic wavefield attributes are reliably determined. For the investigated synthetic data example these indicators permit to assume that the kinematic wavefield attributes are well estimated and, therefore, useful for further applications.

In Figures 6.9 – 6.16 the attributes associated with the inline and crossline stack section (see Figure 6.4) are displayed. For all attribute sections, I applied a coherence mask. This means that each value of an attribute section associated with a semblance value below 0.2 is set to a value which is slightly smaller than the smallest value appearing in the entire section. In this way, only these attribute values are displayed which are really associated with a reflection event. Figure 6.9 shows the azimuth angle α_0. From a theoretical point of view, all attribute values should vary smoothly along a reflection event as discontinuities along traveltimes associated with one reflector are impossible. In fact, also α_0 varies smoothly along the reflection events. The jumps from $+180°$ to $-180°$ in α_0 (i. e. from black to white colors) are no discontinuities but wrap-around effects. The polar angle β_0 is depicted in Figure 6.10.

For the reflection events appearing at the shortest ZO traveltimes (i. e. the uppermost events in the sections), β_0 directly represents the dip of the uppermost reflector in the subsurface at the reflection point of the associated ZO ray. In the inline section of Figure 6.10 a problem becomes obvious which has not been solved by the proposed implementation. At the inline coordinate of $\approx 5000\,\mathrm{m}$ and a traveltime of $\approx 1.8\,\mathrm{s}$, there are two intersecting reflection events. For these ZO samples, there are in fact two associated sets of attributes. However, during the search one event is preferred, i. e., such conflicting dip situations are not accounted for. Mann (2002) gives a solution for the conflicting dip problem for the 2D CRS stack, where, during the search, several coherence maxima associated with one ZO sample are considered. A similar approach is, in principle, also feasible for the 3D CRS stack but will not be discussed here. In Figures 6.11 and 6.12 the inverse of the diagonal elements of the NIP wave curvature matrix $\tilde{k}_{\mathrm{NIP,00}}$ and $\tilde{k}_{\mathrm{NIP,11}}$ are displayed, i. e. the radii of the curvatures of wavefronts. For a plane reflector below a constant-velocity medium, these values directly represent the depth of reflectors. For the uppermost events in the sections of Figures 6.11 and 6.12, the inverse of $\tilde{k}_{\mathrm{NIP,00}}$ and $\tilde{k}_{\mathrm{NIP,11}}$ are the distance of the NIP and the coinciding shot/receiver point of the associated ZO ray. All other values give at least a rough estimate of the reflector depths, which can be seen when comparing the reflector depths of the model (see Figure 6.1 and 6.2) and the values of $\tilde{k}_{\mathrm{NIP,00}}$ and $\tilde{k}_{\mathrm{NIP,11}}$. The non-diagonal element $\tilde{k}_{\mathrm{NIP,01}}$ of the NIP curvature matrix is shown in Figure 6.13. The values of $\tilde{k}_{\mathrm{NIP,01}}$ are close to zero for almost all ZO samples. Theoretically, these values should be exactly zero for the uppermost event as the velocity between the first reflector and measurement surface in the used model is constant and the NIP wavefront has, thus, a rotational symmetry. In Figure 6.13 the values of $\tilde{k}_{\mathrm{NIP,01}}$ associated with the uppermost reflection event deviate only slightly from the predicted values. The elements $\tilde{k}_{\mathrm{N,00}}$, $\tilde{k}_{\mathrm{N,11}}$, and $\tilde{k}_{\mathrm{N,01}}$ of the normal wave curvature matrix are depicted in Figures 6.14, 6.15, and 6.16. The values of $\tilde{k}_{\mathrm{N,00}}$, and $\tilde{k}_{\mathrm{N,11}}$ are directly related to the curvatures of the ZO reflection events in Figure 6.4. For the uppermost events, $\tilde{k}_{N,00}$, $\tilde{k}_{N,11}$, and $\tilde{k}_{N,01}$ can be used together with $\tilde{k}_{NIP,00}$, $\tilde{k}_{NIP,11}$, and $\tilde{k}_{NIP,01}$ to calculate the curvature of the first reflector in the subsurface at the NIP of the associated ZO ray. Furthermore, in a layer-based velocity inversion, as proposed in Höcht et al. (2003), the determined curvature values of the normal wave can be utilised to estimate the local curvatures of reflectors below the first reflector.

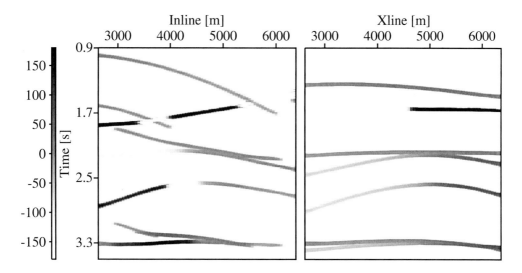

Figure 6.9: Azimuth angles in degrees associated with the reflection events of the ZO sections shown in Figure 6.4.

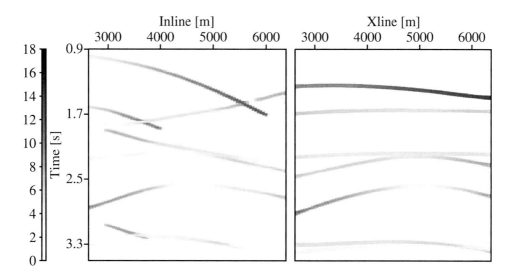

Figure 6.10: Polar angles in degrees associated with the reflection events of the ZO sections shown in Figure 6.4.

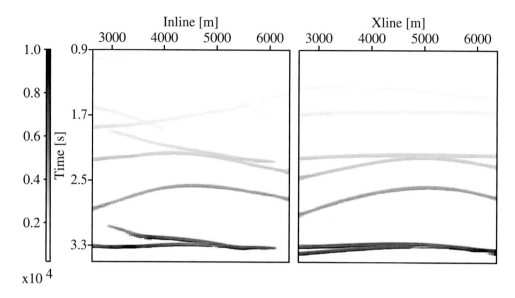

Figure 6.11: Inverse of the element $\tilde{k}_{\text{NIP,00}}$ of the NIP wave curvature matrix in m associated with the reflection events of the ZO sections shown in Figure 6.4.

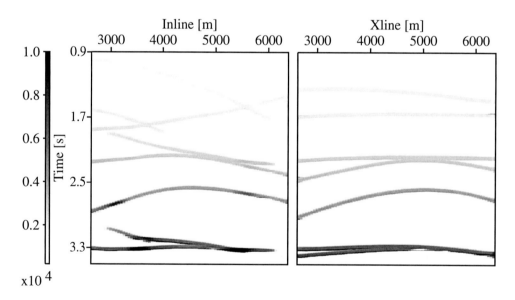

Figure 6.12: Inverse of the element $\tilde{k}_{\text{NIP,11}}$ of the NIP wave curvature matrix in m associated with the reflection events of the ZO sections shown in Figure 6.4.

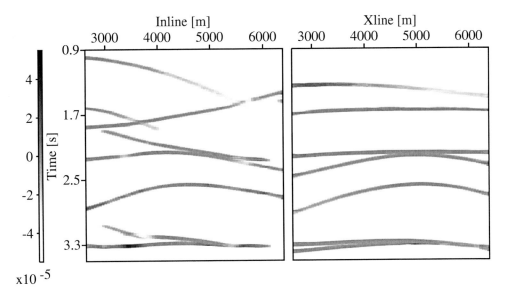

Figure 6.13: Element $\tilde{k}_{\text{NIP},01}$ of the NIP wave curvature matrix in 1/m associated with the reflection events of the ZO sections shown in Figure 6.4.

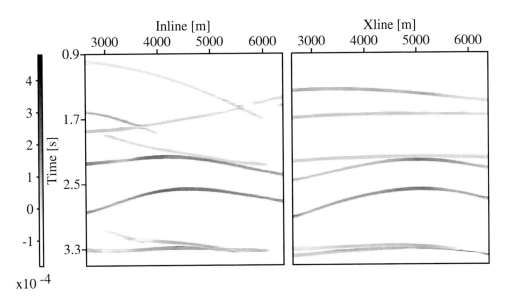

Figure 6.14: Element $\tilde{k}_{\text{N},00}$ of the normal wave curvature matrix in 1/m associated with the reflection events of the ZO sections shown in Figure 6.4.

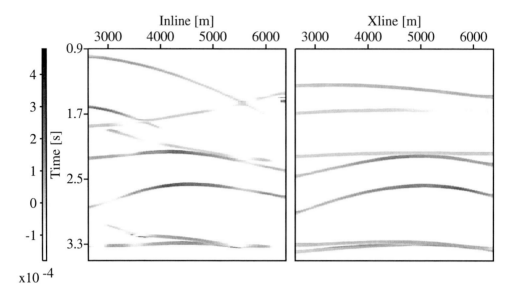

Figure 6.15: Element $\tilde{k}_{N,11}$ of the normal wave curvature matrix in 1/m associated with the reflection events of the ZO sections shown in Figure 6.4.

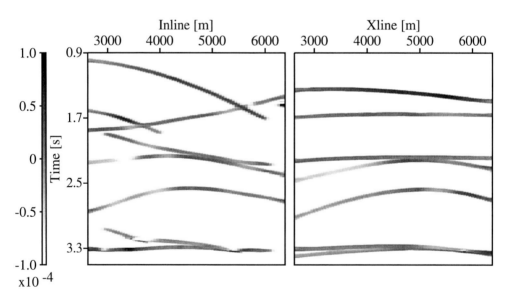

Figure 6.16: Element $\tilde{k}_{N,01}$ of the normal wave curvature matrix in 1/m associated with the reflection events of the ZO sections shown in Figure 6.4.

Chapter 7

Application of the 3D common-reflection-surface stack to real marine data

In order to check the functionality and performance of the CRS search algorithms on complex seismic data, the 3D CRS stack was applied to a real marine dataset. The analysed dataset was provided by courtesy of *WesternGeco*, USA. The data processing was performed in collaboration with *OPERA*, France.

The image quality of the 3D CRS stack results were evaluated by migrating these to depth and comparing the obtained images with results from a 3D prestack depth migration. For both, the poststack and prestack depth migration, the same velocity model constructed by migration velocity analysis was used. In practice, the 3D CRS stack would not be performed together with migration velocity analysis followed by a poststack depth migration to finally obtain a depth image. The comparison is, however, insofar of practical interest as the 3D CRS stack results can be used to establish an alternative way of processing (see Chapter 4). This is possible because the kinematic wavefield attributes obtained during the CRS processing are of use to estimate a velocity model for migration (Duveneck, 2003, 2004). Thus, one can perform the 3D CRS stack, determine the velocity model using the 3D CRS attributes, and then migrate the 3D CRS stack into depth. For 2D datasets this way of processing has already been shown to be successful (Mann et al., 2003; Hertweck, 2004).

7.1 Prestack data

The data under investigation were acquired over a salt body structure embedded in a sedimentary environment. The acquisition was realised by a single vessel towing a multi-streamer. Therefore, a certain azimuthal fold in the CS and, consequently, in CMP volumes has been reached. *TOTAL* as well as *WesternGeco*, who acquired the data, carried out the preprocessing of the data. All subsequently presented results refer to these preprocessed data.

Midpoint and offset geometry	
Number of CMP bins in inline direction	$1619\dots1778$
Number of CMP bins in crossline direction	31
CMP fold	$1\dots93$
Range of Cartesian CMP bin coordinate in inline direction	$\approx0\dots20220\,\text{m}$
Range of Cartesian CMP bin coordinate in crossline direction	$\approx0\dots600\,\text{m}$
CMP bin interval in inline direction	$12.5\,\text{m}$
CMP bin interval in crossline direction	$20\,\text{m}$
Offset range	$289\dots6004\,\text{m}$

Recording parameters	
Recording time	$13\,\text{s}$
Sampling interval	$4\,\text{ms}$
Dominant frequency range	$10\dots40\,\text{Hz}$

Table 7.1: Acquisition parameters of the dataset for the processed area.

The total size of the dataset amounts to several hundreds of gigabytes. The 3D CRS stack was applied to a subset of these data. In order to make the determination of the CRS parameters as reliable as possible, this data subset has a relatively high fold in the CMP volumes (up to 93 traces). However, there are still CMP volumes with a low fold making the obtained results for some (small) areas questionable. The number of CMP bins in inline direction varies between 1619 and 1778, whereas in the crossline direction the number of CMP bins is 31. The distance between two bins in inline direction is 12.5 m and in crossline direction 20 m. Thus, the extension of the area covered by ZO traces produced by the 3D CRS stack is approximately 20000 m × 600 m. The offsets associated with the traces range from 289 m to 6004 m. The recording time is 13 s.[1] The midpoint-offset geometry of the subset of data as well as the recording parameters are summarised in Table 7.1.

7.2 Processing parameters

The processing parameters have been selected according to a) the acquisition scheme and b) the complexity of the reflection events in the data. In the following the choice of all processing parameters is discussed in detail. For a quick overview the processing parameters are listed in Table 7.2.

7.2.1 General processing parameters

Two general parameters apply for all CRS procecssing steps: firstly, the near-surface velocity is given by the propagation velocity of a compressional wave in water. Therefore, the near-surface

---

[1]The long recording time is due to the attempt to register signals from deep reflectors below the salt body.

Output parameters	
Range of CMP coordinate in inline direction	$\approx 0\ldots 20220\,\mathrm{m}$
Range of CMP coordinate in crossline direction	$\approx 0\ldots 600\,\mathrm{m}$
Simulated ZO traveltimes	$1.6\ldots 12.0\,\mathrm{s}$

General search parameters	
Near-surface velocity	$1500\,\mathrm{m/s}$
Total width of coherence time window	$44\,\mathrm{ms}$

Search parameters in CMP volume	
Number of search parameters	1, then 3 (simultaneous)
Moveout increment for rough search	$32\,\mathrm{ms}$
Number of search refinements	0
Stacking velocity range for all azimuths	$1450\ldots 4500\,\mathrm{m/s}$
Offset aperture in x-direction	$6400\,\mathrm{m}$
Offset aperture in y-direction	$2000\,\mathrm{m}$

Search parameters in ZO volume	
Number of search parameters	2 linear (simultaneous)
Moveout increment for rough search	$20\,\mathrm{ms}$
Number of search refinements	0
Midpoint aperture in inline direction	$100\,\mathrm{m}$
Midpoint aperture in crossline direction	$100\,\mathrm{m}$
Range of angle γ	$-20°\ldots 20°$

Stack parameters	
Offset aperture for each CMP bin in inline direction	$2000\,\mathrm{m}$
Offset aperture for each CMP bin in crossline direction	$2000\,\mathrm{m}$
Midpoint aperture in inline direction	$100\,\mathrm{m}$
Midpoint aperture in crossline direction	$100\,\mathrm{m}$

Table 7.2: Selected processing parameters.

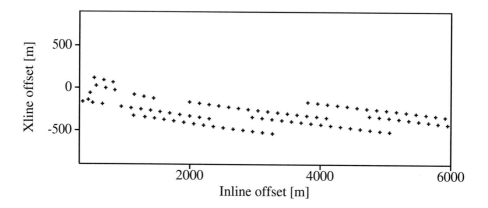

Figure 7.1: Distribution of traces in one CMP volume. Each cross indicates the location of a trace in the inline-crossline offset plane.

velocity is set to 1500 m/s. Secondly, the time window for the semblance analysis is selected according to the dominant signal frequency in the data. It is chosen to be 44 ms which corresponds to a number of eleven time samples.

7.2.2 CMP processing parameters

A full three-parameter search in the CMP volume is not appropriate due to the marine acquisition geometry. Therefore, I decided to determine, in a first step, the stacking velocity in inline direction by a one-parameter search. Subsequently, the stacking velocities in two more directions were calculated which is justified due to the azimuthal distribution of traces. These cover at least a certain range of azimuth angles. Exemplarily, the distribution of traces in the inline-crossline offset plane for one CMP volume is shown in Figure 7.1.

The moveout increment Δt for the stacking velocity analysis described by equations (5.10) has been set to 32 ms. No search refinement (as discussed in Chapter 5) has been performed.[2]

The range of stacking velocities is constrained on the basis of the stacking velocity analyses for a few CMP locations. Due to the small amplitudes of reflection events registered at large traveltimes (larger than 6 s) in the prestack data, the moveouts for these events are difficult to identify. Therefore, I extrapolated the stacking velocities of reflection events for smaller traveltimes linearly to larger traveltimes. This led to a stacking velocity range between 1450 m/s and 4500 m/s which should be sufficient to detect reflection events in the CMP volume.

Based on the inspection of the moveouts of reflection events at several CMP locations, the search aperture in inline direction is set to 6400 m, whereas the aperture in crossline direction is set to

---

[2]The search refinement was not implemented at the time the data were processed.

2000 m. This ensures that the stacking velocity determination is not falsified due to the non-hyperbolicity of reflection events.

7.2.3   ZO processing parameters

In the ZO volumes only a two-parameter search for the linear coefficients of the ZO traveltime formula has been performed. Hence, the search apertures have to be set such that the traveltimes of reflection events can be described locally by a linear function. The search apertures, i. e the maximum inline and crossline midpoint distances from the central trace, are set to 100 m. The quadratic coefficients in the ZO traveltime formula have not been determined and have been set to zero during the entire processing. On the one hand, this reduces the computing time. On the other hand, the number of traces which constructively contribute to the 3D CRS stack is reduced. However, this approach does not alter the concept of the 3D CRS stack, namely the stacking along a hyper-surfaces in the prestack data volume.

The moveout increment Δt for the determination of the linear parameters (equations (5.14)) is set to 40 ms. As for the stacking velocity analysis, no search refinement has been applied.

The dip angle γ which constrains the search of the linear coefficients range between -20° and 20°.

7.2.4   CRS processing parameters

Taking the five determined stacking parameters and setting the coefficients, corresponding to the second traveltime derivatives with respect to the midpoint coordinates, to zero, a five-parameter 3D CRS stack has been performed using equation (5.1). The (hyper-elliptical) apertures of the traveltime hyper-surfaces used for stacking have been selected for each contributing CMP location for the inline and crossline offset to be 2000 m and for the inline and crossline midpoint distances from the associated ZO trace to the central trace to be 100 m.

7.3   Time imaging results

As already mentioned in Chapter 6, during the 3D CRS processing a multitude of output volumes and sections for the different stacks, parameters, etc. are produced. Altogether, these are 25 output volumes. Exemplarily, in Figure 7.2 a part of the ZO volume constructed with the 3D CRS stack is shown. If these volumes are divided into their 31 inline sections for display, a total number of 775 sections would be obtained. However, it would go beyond the scope of this work to present all this information. For this reason, I restrict myself to the results associated with the inline at 300 m, i. e. the inline sections from the middle of the produced volumes. All following results refer to this inline.

The result of the 3D CRS stack is shown in Figures 7.3. The associated coherence section, depicted in Figure 7.4, indicates whether an event has been detected (or not) and, therefore, allows

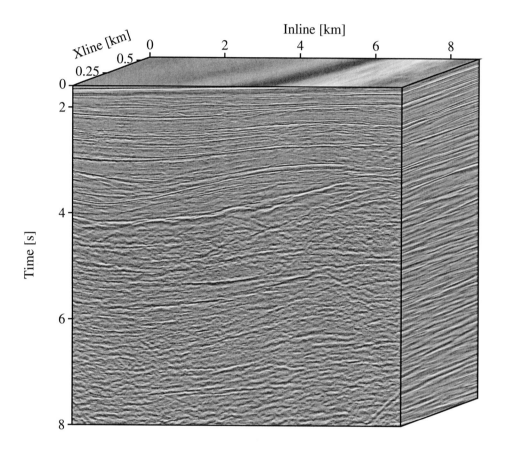

Figure 7.2: Part of a ZO volume produced by the 3D CRS stack.

a kind of quality control. The kinematic wavefield attributes are shown in the following figures: the determined azimuth and polar angles associated with the reflection events are given in Figures 7.5 and 7.6. Figures 7.7 and 7.8 represent the stacking velocity section measured in inline and crossline direction. Figure 7.9 displays the stacking velocities for the azimuth of $45°$. I applied a coherence mask to all attribute sections, i. e., I assigned a value which is slightly smaller than the smallest value appearing in the respective section to all values asscoiated with a semblance smaller than 0.08. In this way, values which are most likely not associated with a reflection event are represented in white. Note that the stacking velocity sections of Figures 7.8 and 7.9 are more speckled than those of Figure 7.7. The reason for this lies in the marine data acquisition geometry which does not allow a stable stacking velocity determination except for the stacking velocities associated with the inline. Therefore, Figures 7.8 and 7.9 provide only a trend and should be treated with care when used for interpretation or further processing.

The 3D CRS stack (Figure 7.3) reveals events between ≈ 1.6 s and ≈ 4.5 s associated with reflections from sedimentary structures. The reflections from the top of the salt body are not completely discernible as these reflections are obscured by several diffraction events. These occur due to the roughness of the top salt and, most likely, due to a system of faults. The diffraction events are easy to identify by their high angle values in Figure 7.6. Most of the polar angles corresponding to reflections from the sedimentary structures above the salt in Figure 7.6 have values between $0°$ and $5°$. These values indicate that the associated reflectors in depth dip only slightly. In Figure 7.3 reflection events from the bottom of the salt body are partly visible and should not be mistaken for a multiple reflection event. The latter was recorded shortly after the reflection event associated with the bottom salt. It can be identified by the low stacking velocities indicated by the arrow in the stacking velocity section (see Figure 7.7). In addition, two more multiples can be identified at traveltimes larger than 7 s by means of their low stacking velocity. After 8.5 s of traveltime, all sections are truncated as there were no reflection events visible at larger traveltimes.

I extracted two parts of Figure 7.3 and enlarged them for comparison with the corresponding parts of the ZO section constructed by the three-parameter CMP stack. The subsets of the 3D CRS stack are shown in Figures 7.10 and 7.12. Those of the 3D CMP stack are displayed in Figures 7.11 and 7.13. The overall impression is that the image quality in the CRS stack section is better than in the CMP stack section. This can be attributed to the increased number of traces which were involved in the construction of the CRS stack sections. The number of traces which contributed to the CRS processing ranges from 500 to 600. This is approximately eight times more than for the CMP stack section which leads to a better signal-to-noise ratio in the CRS stack section. Thus, reflection events are more continuous in the CRS stack section and easier to track. This is an important fact with respect to the picking of events and their associated kinematic wavefield attributes, such as, for instance, the stacking velocities. Increasing the apertures did not lead to an improvement of the image quality. As already mentioned in Chapter 5, if the apertures for the CRS stack are too large, the CRS operator deviates from the true reflection events. Therefore, small-scale features of reflection events in the constructed ZO volume would be lost or, even worse, entire reflection events would be destroyed.

In order to check whether appropriate CRS apertures were chosen, a moveout correction to ZO traveltime was done. As discussed in Chapters 5 and 6, in case the events are flat after moveout correction, the CRS apertures are selected properly. The upper part of Figure 7.14 depicts a CRS

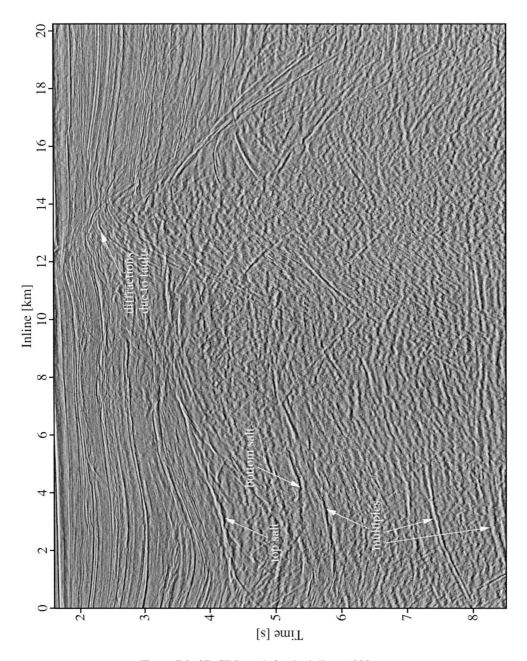

Figure 7.3: 3D CRS stack for the inline at 300m.

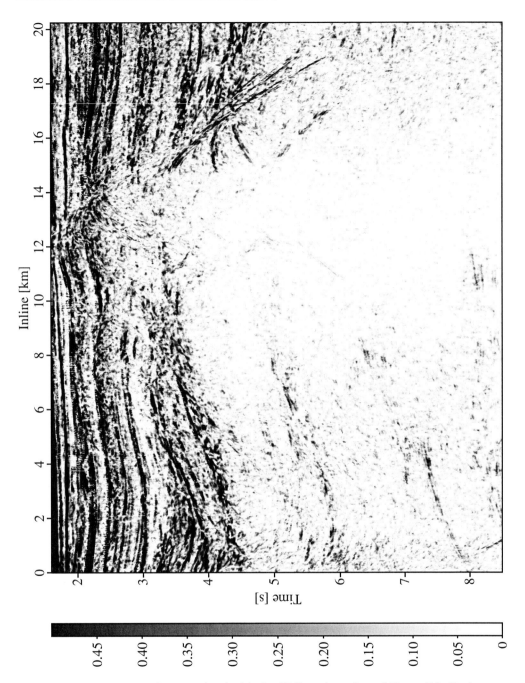

Figure 7.4: Coherence section associated with the CRS stack section of Figure 7.3. Dark parts indicate the detection of reflection events.

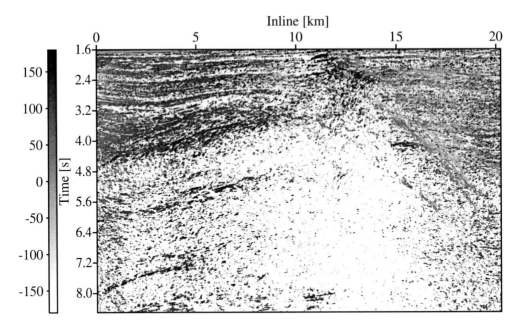

Figure 7.5: Determined azimuth angles in degrees associated with the ZO reflection events in Figure 7.3.

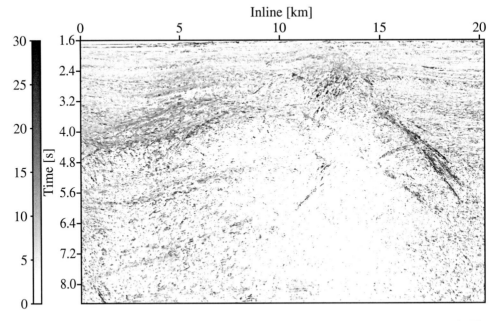

Figure 7.6: Determined polar angles in degrees associated with the ZO reflection events in Figure 7.3.

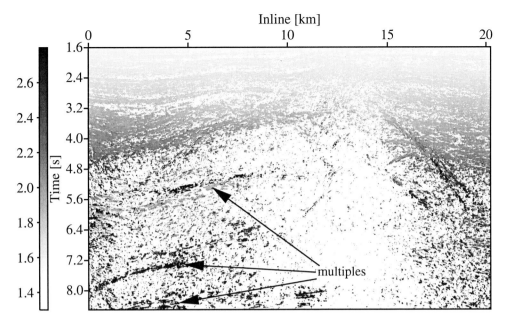

Figure 7.7: Inline stacking velocities in km/s associated with the reflection events of Figure 7.3. The multiples are easier to identify in an appropriately coloured representation of this section.

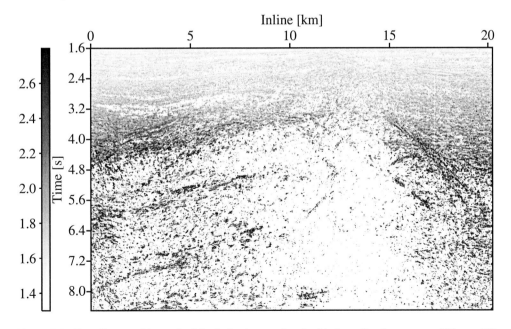

Figure 7.8: Crossline stacking velocities in km/s associated with the reflection events of Figure 7.3.

105

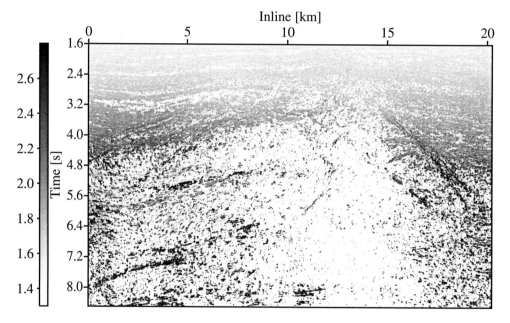

Figure 7.9: Stacking velocities in km/s which can be determined from traces with identical inline and crossline offsets. These are associated with the reflection events of Figure 7.3

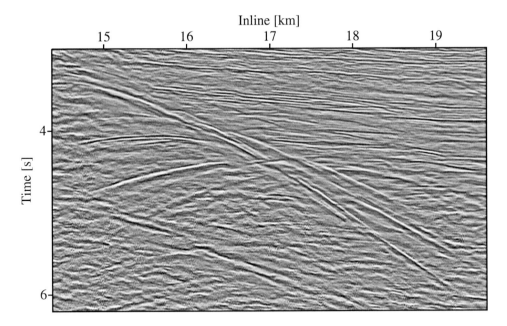

Figure 7.10: Subset of the 3D CRS stack of Figure 7.3.

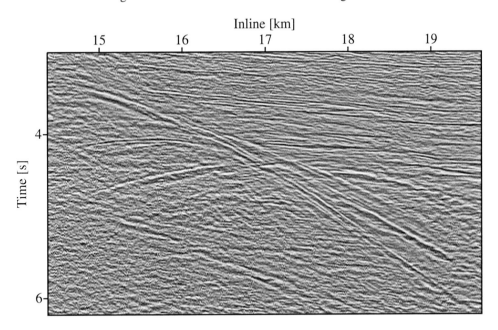

Figure 7.11: Subset of the 3D CMP stack for comparison with Figure 7.10.

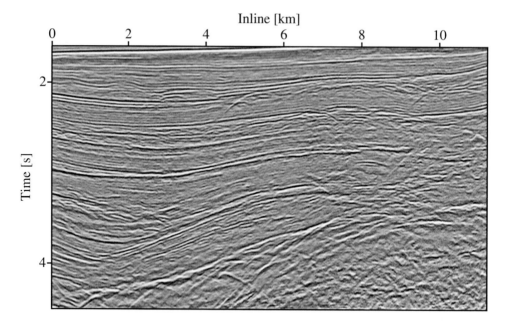

Figure 7.12: Subset of the 3D CRS stack of Figure 7.3.

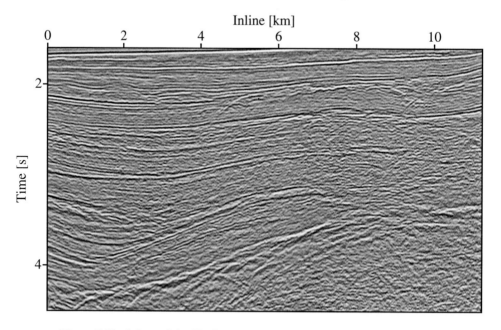

Figure 7.13: Subset of the 3D CMP stack for comparison with Figure 7.12.

gather. It consists of traces from several CMP bins which all contributed to the construction of one 3D CRS stack output trace. The traces are sorted according to their offsets, neglecting their azimuthal positions. For better clarity, only every third trace of the whole CRS gather is shown. In the lower part of Figure 7.14, the same CRS gather is shown, however, after the moveout correction. It demonstrates that the reflection events can be corrected to their ZO traveltime along almost all offsets with the found kinematic wavefield attributes. This confirms that the determined 3D CRS operator is a reasonable approximation for the true reflection traveltime within the chosen apertures.

7.4 Depth imaging results

The time imaging results presented in the previous section may serve for a first interpretation of the seismic data. However, as the subsurface is too complex, the time images are not suitable for detailed structural investigations. Therefore, the ZO volumes constructed with the 3D CRS stack were depth-migrated. This is done by means of 3D Kirchhoff migration. Moreover, the obtained poststack depth-migrated sections are compared with results from prestack depth migration.

In order to keep the comparison as objective as possible, for both, the poststack as well as the prestack migration, the following was done:

- The same 3D Kirchhoff migration code was used which was provided by *OPERA*.

- The traces which contributed to the prestack migration are the same as for the 3D CRS stack to ensure that both approaches use exactly the same information.

- In both cases, the same velocity model was used which was constructed by migration velocity analysis.

Note that the small migration aperture in crossline direction (600 m) prevents both, the poststack as well as the prestack depth migration, from yielding the best possible results.

Figure 7.15 shows the 3D poststack migration result. It represents a vertical slice through the depth-migrated volume which was constructed using the 3D CRS stack volume. The inline and crossline coordinates are the same as for the inline of the 3D CRS stack volume in Figure 7.3. That is, the crossline coordinate is fixed at 300 m. In Figure 7.16, the identical slice through the depth-migrated volume is shown but constructed by 3D prestack depth migration.

In both sections, the top of the salt body as well as parts of the bottom salt are clearly visible. In the poststack migration result there is a strong event between 8 km and 12 km in lateral position and 6 km and 7.5 km in depth I attribute to one of the multiple reflections which was identified by its low stacking velocity (see Section 7.3). Below the salt body, there are no identifiable events. Explanations for this observation are the large velocity variation from the sediments to the salt body as well as the presence of an irregular and steeply dipping salt top, both impeding the illumination of subsalt targets and, therefore, the imaging of these. Moreover, the salt body gives rise to surface multiples which further decrease the possibility to discern subsalt structures.

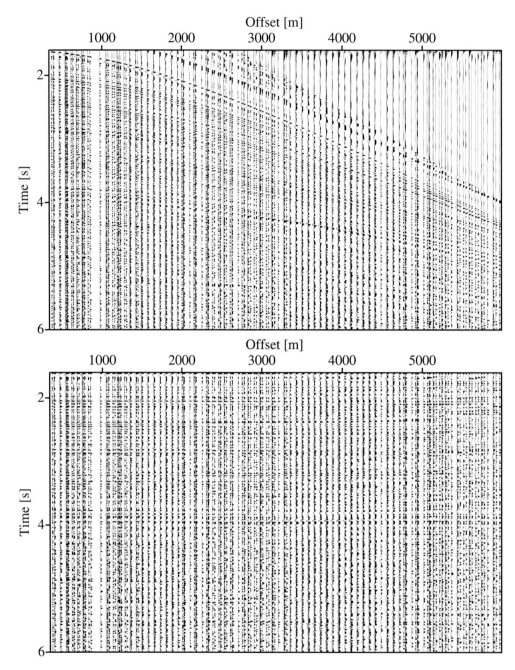

Figure 7.14: Upper part: traces involved in the CRS processing sorted with respect to their offset. Note that traces with different azimuths may be displayed at the same location. Lower part: traces of the upper section after moveout correction.

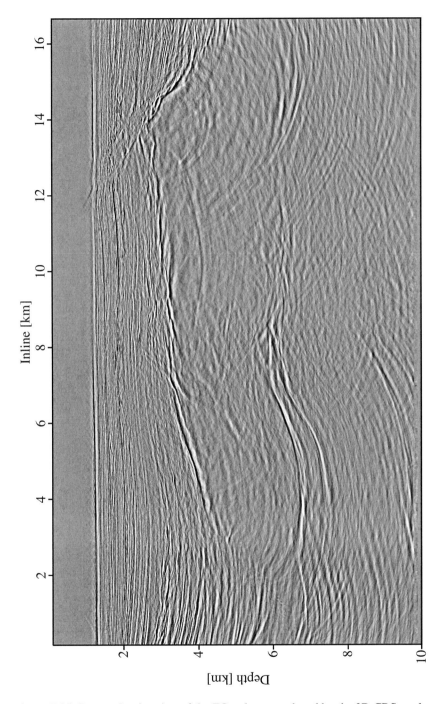

Figure 7.15: Poststack migration of the ZO volume produced by the 3D CRS stack.

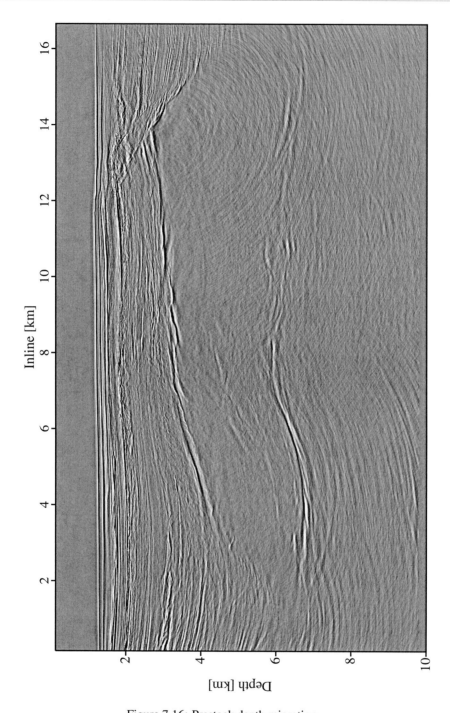

Figure 7.16: Prestack depth migration.

The differences between the poststack and prestack migrated sections are most signficant for the events above and on the flanks of the salt body. In order to better investigate these regions, I enlarged three parts of both sections: Figures 7.17 (poststack) and 7.18 (prestack) reveal a system of faults which cause a large number of diffraction events in the time images. In the prestack migration, the faults are better resolved, especially at the locations indicated by arrows. The reason for this is that the recorded events reflected from such faults become too complex to be described by a second-order traveltime approximation within the selected aperture range. Smaller search and stack apertures for the CRS stack could help in this respect. Moreover, the strong diffraction events render the determination of reflection events during the CRS parameter search difficult. Figures 7.19 (poststack) and 7.20 (prestack) show reflections from virtually flat horizons. The poststack migrated section has a higher vertical resolution which benefits from the high vertical resolution of the CRS stack volume. It resolves the thin layering in the upper part of the subsection much better than the prestack migrated section. In addition, several events in the poststack migrated section are better visible, one of which is emphasized by arrows. A reason for this can be the noise reduction by the 3D CRS stack. One flank of the salt body is shown in Figures 7.21 (poststack) and 7.22 (prestack). In the poststack migrated section, structures along the flank are imaged which are not present in the prestack depth migration. These structures could stem partly from multiple energy which has been suppressed by the prestack migration. However, for the events indicated by arrows, one may assume that these can truely be attributed to geological structures.

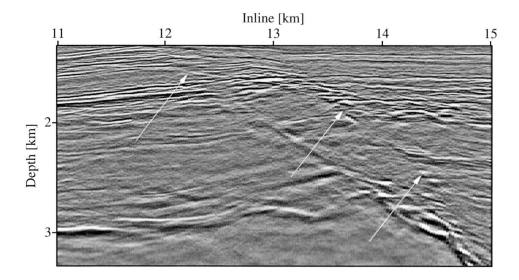

Figure 7.17: Enlarged part of the poststack depth migration: arrows indicate the location of faults which are resolved in the prestack depth-migrated section.

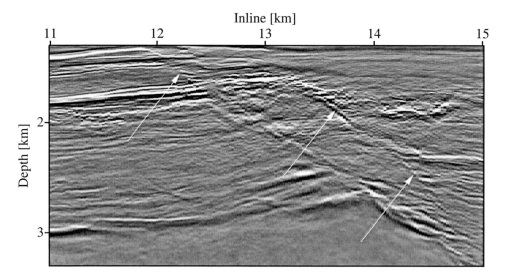

Figure 7.18: Enlarged part of the prestack depth migration: arrows indicate the location of faults.

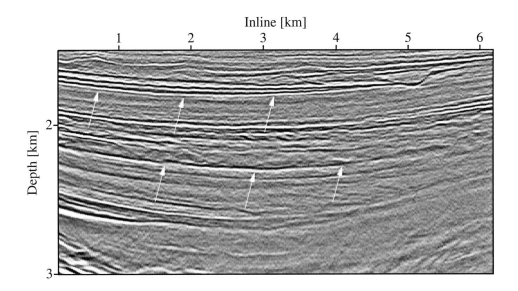

Figure 7.19: Enlarged part of the poststack depth migration: arrows indicate the location of a thin layering and a reflector which are both less resolved in the prestack depth-migrated section.

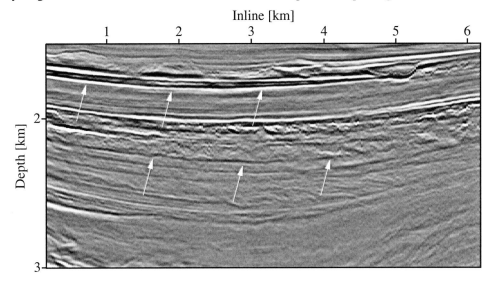

Figure 7.20: Enlarged part of the prestack depth migration: arrows indicate the location of a thin layering and a reflector which are both better resolved in the poststack depth-migrated section.

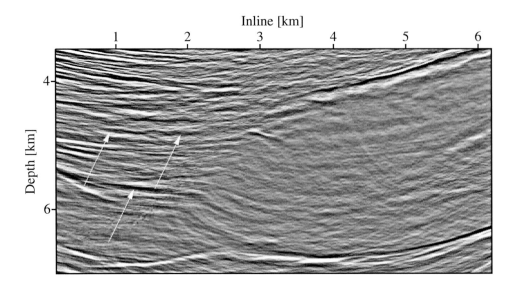

Figure 7.21: Enlarged part of the poststack depth migration: structures at the flanks of the salt body are visible which are missing in the prestack depth-migrated section.

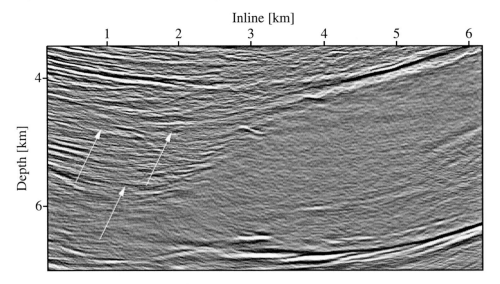

Figure 7.22: Enlarged part of the prestack depth migration: structures at the flanks of the salt body are not as well imaged as in the poststack depth-migrated section.

Chapter 8

Summary and conclusions

In this thesis I studied kinematic wavefield attributes which characterise the propagation directions and curvatures of wavefronts in the framework of the 3D reflection seismic imaging. A main point of the work was the determination and application of the attributes for the 3D CRS stack.

The first part of the thesis dealt with the ray-theoretical foundations of kinematic wavefield attributes, mainly to derive second-order traveltime approximations for reflection events. These approximations are valid for rays with arbitrary starting and end points on a plane measurement surface in the vicinity of a chosen central ray which may also have an arbitrary starting and end point. Two ways were formulated to parameterise the traveltimes: by means of the submatrices of the surface-to-surface propagator matrix and in terms of the kinematic wavefield attributes. In the general case the traveltime formulas are expressed by 14 parameters. In the ZO case the traveltime formulas are described by eight parameters. For multi-coverage seismic data the traveltime approximations locally describe the kinematic reflection response of a reflecting interface in the five-dimensional data hyper-volume. This has two consequences: firstly, by finding a second-order approximation of the kinematic reflection response, the kinematic wavefield attributes can be extracted from the seismic data. The near-surface velocity in the vicinity of the emerging location of the wavefronts is the only quantity of the subsurface which has to be known in advance for this purpose. Secondly, the traveltime approximations serve as stacking operators for the 3D CRS stack. Consequently, traces with arbitrary shot/receiver locations may contribute to a single CRS stacked output trace. Therefore, far more traces are involved in the CRS stack than in a CMP stack, where only traces are used which share the same midpoint (or midpoint bin). For the ZO case, the traveltime formula approximates the kinematic reflection response in the vicinity of a ZO sample. The determination of the 3D CRS operators involves a global non-linear eight-parameter optimisation problem for each ZO sample. This optimisation is, when applied for all parameters simultaneously, completely impractical with regard to the computing time. For this reason, in the second part of the thesis, search algorithms were devised which allow a reasonable computing time and an accurate determination of the parameters. The latter is not only important in terms of the CRS stack but also for the further applications of the kinematic wavefield attributes, such as the determination of a velocity model. In the last part of the work the search algorithms and the 3D CRS stack were applied to a synthetic land dataset and a real marine dataset. In this way the

117

functionality and performance of the proposed 3D CRS processing was validated.

For the investigated real data example it was shown that the poststack depth migration of ZO volumes constructed by means of the 3D CRS stack yields a depth image which is competitive to the depth image obtained by 3D prestack depth migration. This observation applies also in case of complex subsurface structures. For some parts of the depth image the poststack depth migration result was even better which was attributed to the high vertical resolution and the noise suppression of the 3D CRS stack. For the imaging of complex structures it is commonly stated that 3D prestack depth migration is the best solution. However, the presented comparisons let me come to the conclusion that the 3D CRS stack followed by poststack depth migration can be an alternative to 3D prestack depth migration. This conclusion is even more important for data of low quality due to a high noise level. As the 3D CRS stack uses far more traces for the construction of a ZO trace than, for instance, the CMP stack, the 3D CRS stack has the potential to provide, even in such situations, a high signal-to-noise ratio. This simplifies the identification of events which is not only important for the interpretation of seismic images but also for the velocity model building process which often relies on the accurate picking of events.

The 3D CRS stack result is not the final product of the 3D CRS processing. The extraction of the kinematic wavefield attributes from the seismic data and their applications are also important issues. All these applications serve to improve the quality of seismic images. By means of the kinematic wavefield attributes, an entire imaging workflow can be established which involves the 3D CRS stack, the construction of a velocity model using the attributes, and a poststack depth migration using this velocity model. In this way, the 3D CRS processing is also a real alternative with respect to the computing time because not only the expensive 3D prestack depth migration but also the migration velocity analysis can be circumvented. For the 2D case, the CRS based workflow has already been successfully applied. For the 3D case the application of this workflow still remains a future task.

Appendix A

Properties of the propagator matrix $\underline{\mathbf{T}}$

The surface-to-surface propagator concept introduced in Chapter 2 and the laws derived from it can be used for the (mostly approximate) solution of a number of ray-theoretical problems. Among these are the calculation of paraxial traveltimes (as shown in Chapter 2), the computation of geometrical spreading factors (Hubral, 1983) for true-amplitude migration, the analysis of focus phenomena (Bortfeld and Kemper, 1991), or the determination of Fresnel zones (Hubral et al., 1993; Schleicher et al., 1997). For the treatment of these problems, the knowledge of properties of the surface-to-surface propagator matrix $\underline{\mathbf{T}}$ is of help. In this chapter, I review some of these properties.

A.1 Symplecticity

For the second derivatives of the traveltimes, I have to demand (see Chapter 2) that the order of differentiation is immaterial. This leads to symmetries of $\underline{\mathbf{T}}$, as this matrix is related to second traveltime derivatives. These symmetries are denoted as the symplecticity. This property can be expressed by the inverse matrix of $\underline{\mathbf{T}}$:

$$\underline{\mathbf{T}}^{-1} = \begin{pmatrix} \mathbf{A} & \mathbf{B} \\ \mathbf{C} & \mathbf{D} \end{pmatrix}^{-1} = \begin{pmatrix} \mathbf{D}^{\mathrm{T}} & -\mathbf{B}^{\mathrm{T}} \\ -\mathbf{C}^{\mathrm{T}} & \mathbf{A}^{\mathrm{T}} \end{pmatrix}. \tag{A.1}$$

The identity $\underline{\mathbf{T}}^{-1}\,\underline{\mathbf{T}} = \underline{\mathbf{I}}$ then yields the following conditions which $\underline{\mathbf{T}}$ has to fulfil:

$$\mathbf{B}^{-1}\mathbf{A} = \left(\mathbf{B}^{-1}\mathbf{A} \right)^{\mathbf{T}} \tag{A.2a}$$

$$\mathbf{D}\mathbf{B}^{-1} = \left(\mathbf{D}\mathbf{B}^{-1} \right)^{\mathbf{T}} \tag{A.2b}$$

$$\mathbf{A}^{\mathrm{T}}\mathbf{D} - \mathbf{C}^{\mathrm{T}}\mathbf{B} = \mathbf{I}. \tag{A.2c}$$

A.2 Reverse ray

If I interchange starting and end points of a paraxial ray, i. e., if I put the starting point at G and the end point at S (see Figure 2.3), the starting point is defined by the former end point vector and vice versa. The slowness vector projections at the starting and end point of the reversed ray from G to S have the same norm as the ray from S to G but the opposite direction. Using this information, the elements of the propagator matrix of the reversed rays $\underline{\mathbf{T}}^*$ can be related to $\underline{\mathbf{T}}$ by (see, e. g. Hubral, 1983)

$$\underline{\mathbf{T}}^* = \begin{pmatrix} \mathbf{D}^{\mathrm{T}} & \mathbf{B}^{\mathrm{T}} \\ \mathbf{C}^{\mathrm{T}} & \mathbf{A}^{\mathrm{T}} \end{pmatrix} . \tag{A.3}$$

In the ZO case, when the central ray is a normal ray, the matrices $\underline{\mathbf{T}}^*$ and $\underline{\mathbf{T}}$ are identical. Comparing both matrices yields

$$\mathbf{D}^{\mathrm{T}} = \mathbf{A} . \tag{A.4}$$

Together with the symplecticity, this reduces the number of independent submatrices of $\underline{\mathbf{T}}$ to two for the ZO case. Moreover, it becomes obvious that the submatrices $\mathbf{B}$ and $\mathbf{C}$ are symmetric in the ZO case.

A.3 Chain rule

If the central ray is decomposed into two ray branches, the propagator matrix corresponding to the entire ray can be written in terms of the propagator matrices corresponding to the two individual ray branches. For example, a reflected ray can be decomposed into a branch from the starting point to the reflection point and from the reflection point to the end point. Using the fact that the slowness vector component tangent to an interface is invariant (Snell's law), it follows that (see, e. g., Hubral et al., 1992b)

$$\underline{\mathbf{T}} = \underline{\mathbf{T}}_2 \underline{\mathbf{T}}_1 . \tag{A.5}$$

The matrix $\underline{\mathbf{T}}_1$ denotes the propagator matrix of the first ray branch and the matrix $\underline{\mathbf{T}}_2$ corresponds to the second ray branch. As the matrices $\underline{\mathbf{T}}_1$ and $\underline{\mathbf{T}}_2$ can be further decomposed, $\underline{\mathbf{T}}$ can be written as a product of many ray-segment propagator matrices.

If the propagator matrix from the starting point to the reflection point of a normal ray is denoted by $\underline{\mathbf{T}}_0$ (as in Chapter 2), then the total propagator matrix $\underline{\mathbf{T}}$ can be formulated due to equations (A.3) and (A.5) as

$$\underline{\mathbf{T}} = \underline{\mathbf{T}}_0^* \underline{\mathbf{T}}_0 . \tag{A.6}$$

The matrix product shows that

$$\underline{\mathbf{T}} = \begin{pmatrix} \mathbf{A} & \mathbf{B} \\ \mathbf{C} & \mathbf{D} \end{pmatrix} = \begin{pmatrix} \mathbf{D}_0{}^{\mathrm{T}}\mathbf{A}_0 + \mathbf{B}_0{}^{\mathrm{T}}\mathbf{C}_0 & \mathbf{D}_0{}^{\mathrm{T}}\mathbf{B}_0 + \mathbf{B}_0{}^{\mathrm{T}}\mathbf{D}_0 \\ \mathbf{C}_0{}^{\mathrm{T}}\mathbf{A}_0 + \mathbf{A}_0{}^{\mathrm{T}}\mathbf{C}_0 & \mathbf{C}_0{}^{\mathrm{T}}\mathbf{B}_0 + \mathbf{A}_0{}^{\mathrm{T}}\mathbf{D}_0 \end{pmatrix} . \tag{A.7}$$

Note that equation (A.7) restates equation (A.4).

Appendix B

Slowness vectors at curved surfaces

The aim of this appendix is to demonstrate that the third components of position and slowness vectors do not contribute to a second-order traveltime approximation for paraxial rays if starting and end points are located on curved surfaces. Thus, the third components in the dot products of Hamilton's equation (2.14) can be neglected for paraxial rays. The following consideration refer to the particular situation considered in Section 2.5, where the end points of the central and paraxial ray are on a reflecting interface and the central ray is a normal ray (see Figure 2.4). However, these derivations are valid in general, that is, for any curved surface on which starting and end points of paraxial rays are located.

For the problem description, a local Cartesian coordinate system like the primed coordinate system defined in Section 2.5 is best suitable. Thus, all quantities in the following refer to the primed coordinate system.

I identify the point R on the reflecting interface in Figure 2.4 with the end point of a paraxial ray. According to equation (2.12), an infinitesimal dislocation of R leads to a small change in traveltime. This is given by

$$dt_R = \hat{\mathbf{p}}'_R \cdot d\hat{\mathbf{r}}'_R = \hat{\mathbf{p}}'_{R,T} \cdot d\hat{\mathbf{r}}'_R,$$ (B.1)

where $\hat{\mathbf{p}}'_{R,T}$ is the orthogonal projection of the slowness vector $\hat{\mathbf{p}}'_R$ onto the plane which is tangent to the reflector at the paraxial reflection point (see Figure 2.4). The vector $d\hat{\mathbf{r}}'_R$ describes an infinitesimal dislocation along the reflector and is, therefore, tangent to the reflector. Thus, the contribution from the component of vector $\hat{\mathbf{p}}'_R$ vertical to the reflector in the dot product of equation (B.1) vanishes. When $z' = f(x', y')$ describes the reflection surface in the vicinity of the NIP, i. e. the end point of the central ray, then vector $d\hat{\mathbf{r}}'_R$ is expressed by

$$d\hat{\mathbf{r}}'_R = \begin{pmatrix} d\mathbf{r}'_R \\ \nabla f \cdot d\mathbf{r}'_R \end{pmatrix},$$ (B.2)

where $d\mathbf{r}'_R$ is the two-component vector of $d\hat{\mathbf{r}}'_R$ in the $x'y'$-plane. Correspondingly, the vector $\hat{\mathbf{p}}'_{R,T}$ is expressed by

$$\hat{\mathbf{p}}'_{R,T} = \begin{pmatrix} \mathbf{p}'_R \\ \nabla f \cdot \mathbf{p}'_R \end{pmatrix},$$ (B.3)

where $\mathbf{p}'_\mathbf{R}$ is the two-component representation of $\hat{\mathbf{p}}'_{\mathbf{R},\mathbf{T}}$ in the $x'y'$-plane. Due to the choice of the primed coordinate system, there are no constant or linear terms in the series expansion of f. Therefore, the result of $\nabla f \cdot d\mathbf{r}'_\mathbf{R} \, \nabla f \cdot \mathbf{p}'_\mathbf{R}$ is at least of second order. Thus, the third components of position and slowness vectors do not have to be considered in Hamilton's equation (2.14) if a second-order traveltime approximation is sought.

Appendix C

Mixed second traveltime derivatives and kinematic wavefield attributes

In Chapter 3 the inverse of the matrix $\mathbf{B}$ in equation (2.30) representing the mixed second travel-time derivatives has been related to kinematic wavefield attributes. The result of this relationship is given in form of the individual elements of the matrix $\mathbf{E}$. The explanation for the different terms in the following equations can be found in Chapter 3. The elements of $\mathbf{E}$ read:

$$
\begin{aligned}
e_{00} = &\frac{1}{2v_{\mathrm{S}}} \left(\tilde{k}_{\mathrm{S},00}^{\mathrm{CR}} - \tilde{k}_{\mathrm{S},00}^{\mathrm{CMP}} \right) \cos^2 \alpha_{\mathrm{S}} \cos^2 \beta_{\mathrm{S}} + \frac{1}{2v_{\mathrm{G}}} \left(\tilde{k}_{\mathrm{G},00}^{\mathrm{CMP}} - \tilde{k}_{\mathrm{G},00}^{\mathrm{CS}} \right) \cos^2 \alpha_{\mathrm{G}} \cos^2 \beta_{\mathrm{G}} \\
&+ \frac{1}{2v_{\mathrm{S}}} \left(\tilde{k}_{\mathrm{S},11}^{\mathrm{CR}} - \tilde{k}_{\mathrm{S},11}^{\mathrm{CMP}} \right) \sin^2 \alpha_{\mathrm{S}} + \frac{1}{2v_{\mathrm{G}}} \left(\tilde{k}_{\mathrm{G},11}^{\mathrm{CMP}} - \tilde{k}_{\mathrm{G},11}^{\mathrm{CS}} \right) \sin^2 \alpha_{\mathrm{G}} \\
&+ \frac{1}{v_{\mathrm{S}}} \left(\tilde{k}_{\mathrm{S},01}^{\mathrm{CMP}} - \tilde{k}_{\mathrm{S},01}^{\mathrm{CR}} \right) \cos \alpha_{\mathrm{S}} \cos \beta_{\mathrm{S}} \sin \alpha_{\mathrm{S}} + \frac{1}{v_{\mathrm{G}}} \left(\tilde{k}_{\mathrm{G},01}^{\mathrm{CS}} - \tilde{k}_{\mathrm{G},01}^{\mathrm{CMP}} \right) \cos \alpha_{\mathrm{G}} \cos \beta_{\mathrm{G}} \sin \alpha_{\mathrm{G}}
\end{aligned}
\tag{C.1}
$$

$$
\begin{aligned}
e_{01} = &\frac{1}{2v_{\mathrm{S}}} \left(\tilde{k}_{\mathrm{S},00}^{\mathrm{CP}} - \tilde{k}_{\mathrm{S},00}^{\mathrm{CR}} \right) \cos^2 \alpha_{\mathrm{S}} \cos^2 \beta_{\mathrm{S}} + \frac{1}{2v_{\mathrm{G}}} \left(\tilde{k}_{\mathrm{G},00}^{\mathrm{CS}} - \tilde{k}_{\mathrm{G},00}^{\mathrm{CP}} \right) \cos^2 \alpha_{\mathrm{G}} \cos^2 \beta_{\mathrm{G}} \\
&+ \frac{1}{2v_{\mathrm{S}}} \left(\tilde{k}_{\mathrm{S},11}^{\mathrm{CP}} - \tilde{k}_{\mathrm{S},11}^{\mathrm{CR}} \right) \sin^2 \alpha_{\mathrm{S}} + \frac{1}{2v_{\mathrm{G}}} \left(\tilde{k}_{\mathrm{G},11}^{\mathrm{CS}} - \tilde{k}_{\mathrm{G},11}^{\mathrm{CP}} \right) \sin^2 \alpha_{\mathrm{G}} \\
&+ \frac{1}{v_{\mathrm{S}}} \left(\tilde{k}_{\mathrm{S},01}^{\mathrm{CR}} - \tilde{k}_{\mathrm{S},01}^{\mathrm{CP}} \right) \cos \alpha_{\mathrm{S}} \cos \beta_{\mathrm{S}} \sin \alpha_{\mathrm{S}} + \frac{1}{v_{\mathrm{G}}} \left(\tilde{k}_{\mathrm{G},01}^{\mathrm{CP}} - \tilde{k}_{\mathrm{G},01}^{\mathrm{CS}} \right) \cos \alpha_{\mathrm{G}} \cos \beta_{\mathrm{G}} \sin \alpha_{\mathrm{G}}
\end{aligned}
\tag{C.2}
$$

$$
\begin{aligned}
e_{10} = &\frac{1}{2v_{\mathrm{S}}} \left(\tilde{k}_{\mathrm{S},00}^{\mathrm{CP}} - \tilde{k}_{\mathrm{S},00}^{\mathrm{CR}} \right) \cos^2 \alpha_{\mathrm{S}} \cos^2 \beta_{\mathrm{S}} + \frac{1}{2v_{\mathrm{G}}} \left(\tilde{k}_{\mathrm{G},00}^{\mathrm{CS}} - \tilde{k}_{\mathrm{G},00}^{\mathrm{CP}} \right) \cos^2 \alpha_{\mathrm{G}} \cos^2 \beta_{\mathrm{G}} \\
&+ \frac{1}{2v_{\mathrm{S}}} \left(\tilde{k}_{\mathrm{S},11}^{\mathrm{CR}} - \tilde{k}_{\mathrm{S},11}^{\mathrm{CP}} \right) \cos^2 \alpha_{\mathrm{S}} + \frac{1}{2v_{\mathrm{G}}} \left(\tilde{k}_{\mathrm{G},11}^{\mathrm{CP}} - \tilde{k}_{\mathrm{G},11}^{\mathrm{CS}} \right) \cos^2 \alpha_{\mathrm{G}} \\
&+ \frac{1}{2v_{\mathrm{S}}} \left(\tilde{k}_{\mathrm{S},00}^{\mathrm{CR}} - \tilde{k}_{\mathrm{S},00}^{\mathrm{CP}} \right) \cos^2 \beta_{\mathrm{S}} + \frac{1}{2v_{\mathrm{G}}} \left(\tilde{k}_{\mathrm{G},00}^{\mathrm{CP}} - \tilde{k}_{\mathrm{G},00}^{\mathrm{CS}} \right) \cos^2 \beta_{\mathrm{G}} \\
&+ \frac{1}{v_{\mathrm{S}}} \left(\tilde{k}_{\mathrm{S},01}^{\mathrm{CR}} - \tilde{k}_{\mathrm{S},01}^{\mathrm{CP}} \right) \cos \alpha_{\mathrm{S}} \cos \beta_{\mathrm{S}} \sin \alpha_{\mathrm{S}} + \frac{1}{v_{\mathrm{G}}} \left(\tilde{k}_{\mathrm{G},01}^{\mathrm{CP}} - \tilde{k}_{\mathrm{G},01}^{\mathrm{CS}} \right) \cos \alpha_{\mathrm{G}} \cos \beta_{\mathrm{G}} \sin \alpha_{\mathrm{G}} \,,
\end{aligned}
\tag{C.3}
$$

and

$$
\begin{aligned}
e_{11} =& \frac{1}{2v_{\mathrm{S}}} \left(\tilde{k}_{\mathrm{S},00}^{\mathrm{CMP}} - \tilde{k}_{\mathrm{S},00}^{\mathrm{CR}} \right) \cos^2 \alpha_{\mathrm{S}} \cos^2 \beta_{\mathrm{S}} + \frac{1}{2v_{\mathrm{G}}} \left(\tilde{k}_{\mathrm{G},00}^{\mathrm{CS}} - \tilde{k}_{\mathrm{G},00}^{\mathrm{CMP}} \right) \cos^2 \alpha_{\mathrm{G}} \cos^2 \beta_{\mathrm{G}} \\
&+ \frac{1}{2v_{\mathrm{S}}} \left(\tilde{k}_{\mathrm{S},11}^{\mathrm{CR}} - \tilde{k}_{\mathrm{S},11}^{\mathrm{CMP}} \right) \cos^2 \alpha_{\mathrm{S}} + \frac{1}{2v_{\mathrm{G}}} \left(\tilde{k}_{\mathrm{G},11}^{\mathrm{CMP}} - \tilde{k}_{\mathrm{G},11}^{\mathrm{CS}} \right) \cos^2 \alpha_{\mathrm{G}} \\
&+ \frac{1}{2v_{\mathrm{S}}} \left(\tilde{k}_{\mathrm{S},00}^{\mathrm{CR}} - \tilde{k}_{\mathrm{S},00}^{\mathrm{CMP}} \right) \cos^2 \beta_{\mathrm{S}} + \frac{1}{2v_{\mathrm{G}}} \left(\tilde{k}_{\mathrm{G},00}^{\mathrm{CMP}} - \tilde{k}_{\mathrm{G},00}^{\mathrm{CS}} \right) \cos^2 \beta_{\mathrm{G}} \\
&+ \frac{1}{v_{\mathrm{S}}} \left(\tilde{k}_{\mathrm{S},01}^{\mathrm{CR}} - \tilde{k}_{\mathrm{S},01}^{\mathrm{CMP}} \right) \cos \alpha_{\mathrm{S}} \cos \beta_{\mathrm{S}} \sin \alpha_{\mathrm{S}} + \frac{1}{v_{\mathrm{G}}} \left(\tilde{k}_{\mathrm{G},01}^{\mathrm{CMP}} - \tilde{k}_{\mathrm{G},01}^{\mathrm{CS}} \right) \cos \alpha_{\mathrm{G}} \cos \beta_{\mathrm{G}} \sin \alpha_{\mathrm{G}} .
\end{aligned}
\tag{C.4}
$$

Appendix D

The normal-incidence point wave theorem

It has been well-known for many years that the rays involved in a CMP experiment in general do not share the same point on the reflector (see, e. g. Levin, 1971). This is true only for horizontally layered media with constant velocities in each layer.

The aim of this appendix is to show that the rays in the vicinity of a normal central ray, which belong to a CMP experiment (see Subsection 3.5.2), can be associated with the NIP wave. As the NIP wave can be initiated by a point source at the NIP of the central ray on the reflector, it is necessary to prove that the reflection points of all CMP rays in the paraxial approximation are identical with the NIP.

Summing equations (2.33b) and (2.34b) and solving for vector $\mathbf{r}'_\mathbf{R}$ yields

$$\mathbf{r}'_\mathbf{R} \approx \left(\mathbf{A_0} - \mathbf{B_0} \mathbf{D_0^{-1}} \mathbf{C_0} \right) \frac{1}{2} \left(\mathbf{r_G} + \mathbf{r}'_\mathbf{S} \right) = \left(\mathbf{A_0} - \mathbf{B_0} \mathbf{D_0^{-1}} \mathbf{C_0} \right) \Delta \mathbf{m} . \tag{D.1}$$

Vector $\mathbf{r}'_\mathbf{R}$ describes the locations of the reflection points of the paraxial rays in the vicinity of the NIP. From equation (D.1) it becomes obvious that the location of the reflection points depends on the midpoints of the central and paraxial rays only. In case of the CMP experiment, where all involved rays have the identical midpoint, the reflection point is, therefore,—within the paraxial approximation—the same for all CMP rays, namely the NIP.

Equation (D.1) was derived in Bortfeld (1989). Although not noted there, it represents the NIP wave theorem. It states that all rays of a CMP experiment can be viewed as passing through the NIP as long as one deals with a second-order approximation of traveltime along the CMP rays (Chernjak and Gritsenko, 1979; Hubral and Krey, 1980; Hubral, 1983). As shown in Chapter 2, the paraxial approximation of equation (2.16) leads to a second-order traveltime approximation. Therefore, a NIP wave initiated by a point source at the NIP has the identical wavefront curvatures at the coinciding starting and end point of the central ray as the curvatures of the surface, which is orthogonal to the CMP rays at this point.

Appendix E

Equations for a 3D attribute-based time migration

Time migration is performed to provide a more realistic image of the subsurface geology than stacked sections (such as, for instance, the ZO volume constructed by a CRS or CMP stack) in the time domain, particularly when the recorded wavefield contains plenty of diffracted energy. In this appendix the formulas for a 3D time migration, based on the kinematic wavefield attributes, are presented. This time migration is the extension of the 2D approach discussed in Mann (2002). The formulas are introduced without giving any theoretical background on time migration. For a comprehensive discussion on time migration the reader is referred to Hubral and Krey (1980).

The attribute-based time migration is a Kirchhoff-type approach which involves the summation of recorded signals into the apex of traveltime surfaces associated with a diffractor (Hubral and Krey, 1980). A diffractor can be defined as a reflector segment with infinite curvature in each spatial direction. In terms of the normal wave experiment (see Chapter 3) this means that the normal wave shrinks to a point at the reflection point of the central ZO ray. Thus, it emerges with exactly the same wavefront curvatures at the ZO location on the measurement surface as the wavefront of the NIP wave, i. e.

$$\tilde{\mathbf{K}}_{\mathrm{N}} = \tilde{\mathbf{K}}_{\mathrm{NIP}} . \tag{E.1}$$

Inserting this condition into equation (3.71), yields

$$t_{\mathrm{diff}}^2(\Delta\mathbf{m},\mathbf{h}) \approx \left(t_0 + 2\mathbf{p_0}\cdot\Delta\mathbf{m}\right)^2 + \frac{2t_0}{v_0}\Delta\mathbf{m}\cdot\mathbf{R}\,\tilde{\mathbf{K}}_{\mathrm{NIP}}\,\mathbf{R}^{\mathrm{T}}\Delta\mathbf{m} + \frac{2t_0}{v_0}\mathbf{h}\cdot\mathbf{R}\,\tilde{\mathbf{K}}_{\mathrm{NIP}}\,\mathbf{R}^{\mathrm{T}}\mathbf{h} . \tag{E.2}$$

Equation (E.2) describes an approximation of the kinematic response of a diffraction point in a five-dimensional data hyper-volume with its apex lying in the ZO volume. In this volume equation (E.2) reduces to

$$t_{\mathrm{diff,ZO}}^2(\Delta\mathbf{m}) \approx \left(t_0 + 2\mathbf{p_0}\cdot\Delta\mathbf{m}\right)^2 + \frac{2t_0}{v_0}\Delta\mathbf{m}\cdot\mathbf{R}\,\tilde{\mathbf{K}}_{\mathrm{NIP}}\,\mathbf{R}^{\mathrm{T}}\Delta\mathbf{m} , \tag{E.3}$$

which is simply obtained by inserting $\mathbf{h} = (0,0)^T$ into equation (E.2). The midpoint location of the apex can be evaluated with equation (E.3) using the conditions

$$\frac{\partial t_{\text{diff,ZO}}}{\partial \Delta m_x} = 0 \qquad \text{and} \qquad \frac{\partial t_{\text{diff,ZO}}}{\partial \Delta m_y} = 0, \tag{E.4}$$

and is given by

$$\Delta \mathbf{m}_{\text{apex}} = -2t_0 \left(4\mathbf{p_0 p_0}^T + \mathbf{R}\, \tilde{\mathbf{K}}_{\text{NIP}}\, \mathbf{R}^T \right)^{-1} \mathbf{p_0}. \tag{E.5}$$

Substituting expression (E.5) for $\Delta \mathbf{m}$ in equation (E.3) yields the corresponding ZO traveltime at the apex location.

Performing the 3D CRS stack as described in Chapter 5 but using the diffraction operator (E.2) and placing the stack result into the apex location yields the attribute-based time-migrated image. Note that this can be easily performed as soon as the 3D CRS stack is conducted because it makes use of exactly the same kinematic wavefield attributes as the CRS stack.

List of Figures

Chapter 6 – Test of the 3D common-reflection-surface stack implementation on synthetic data 79

List of Tables

References

Al-Yahya, K. (1989). Velocity analysis by iterative profile migration. *Geophysics*, 54(6):718–729.

Bergler, S., Hubral, P., Marchetti, P., Cristini, A., and Cardone, G. (2002). 3D common-reflection-surface stack and kinematic wavefield attributes. *The Leading Edge*, 21(10):1010–1015.

Berkhout, A. (1997a). Pushing the limits of seismic imaging, Part I: Prestack migration in terms of double dynamic focusing. *Geophysics*, 62(3):937–953.

Berkhout, A. (1997b). Pushing the limits of seismic imaging, Part II: Integration of prestack migration, velocity estimation, and AVO analysis. *Geophysics*, 62(3):954–969.

Billete, F., Le Bégat, S., Podvin, P., and Lambaré, G. (2003). Practical aspects and applications of 2D stereotomography. *Geophysics*, 68(3):1008–1021.

Billette, F. and Lambaré, G. (1998). Velocity macromodel estimation from seismic reflection data by stereotomography. *Geophys. J. Int.*, 135:671–690.

Biondi, B. (2003). *3-D Seismic Imaging*. http://sepwww.stanford.edu/sep/biondo/Lectures/, Stanford University.

Bleistein, N. (1984). *Mathematical methods for wave phenomena*. Academic Press Inc., Orlando.

Bolte, J. F. B. (2003). *Estimation of focusing operators using the Common Focal Point method*. PhD thesis, Delft University of Technology, The Netherlands.

Bortfeld, R. (1989). Geometrical ray theory: Rays and traveltimes in seismic systems (second-order approximations of the traveltimes). *Geophysics*, 54(3):342–349.

Bortfeld, R. and Kemper, M. (1991). Geometrical ray theory: Line foci and point foci in the anterior surface of seismic systems (second-order approximation of the traveltimes). *Geophysics*, 56(6):806–811.

Červený, V. (2001). *Seismic Ray Theory*. Cambridge University Press.

Chernjak, V. S. and Gritsenko, S. A. (1979). Interpretation of effective common depth point parameters for a spatial system of homogeneous beds with curved boundaries. *Soviet Geology and Geophysics*, 20(12):91–98.

Cox, B. E., Winthaegen, P. L. A., Verschuur, D. J., and Roy-Chowdhury, K. (2001). Common Focus Point velocity estimation for laterally varying velocities. *First Break*, 19(2):75–85.

Cristini, A., Cardone, G., and Marchetti, P. (2002). 3D Common Reflection Surface Stack for land data – real data example. In *Extended Abstracts*. 64th Mtg., Eur. Assn. Geosci. Eng. Session B-015.

de Bazelaire, E. (1988). Normal moveout revisited: Inhomogeneous media and curved interfaces. *Geophysics*, 53(2):143–157.

de Bazelaire, E. and Viallix, J. R. (1994). Normal moveout in focus. *Geophys. Prosp.*, 42:477–499.

Dix, C. H. (1955). Seismic velocities from surface measurements. *Geophysics*, 20(1):68–86.

Douze, E. J. and Laster, S. J. (1979). Statistics of semblance. *Geophysics*, 44(12):1999–2003.

Duveneck, E. (2003). Tomographic velocity model estimation with CRS attributes. In *EAGE/SEG Summer Research Workshop on 'Processing and imaging of seismic data'—Using explicit or implicit velocity model information?, Extended Abstracts Book*, Session T13.

Duveneck, E. (2004). Velocity model estimation with data-derived wavefront attributes. *Geophysics*, 69(1):265–274.

Duveneck, E. and Hubral, P. (2002). Tomographic velocity inversion using kinematic wavefield attributes. In *Expanded Abstracts*, pages 862–865. 72nd Annual Internat. Mtg., Soc. Expl. Geophys.

Gelchinsky, B., Berkovitch, A., and Keydar, S. (1999). Multifocusing homeomorphic imaging Part 1. Basic concepts and formulas. *J. Appl. Geoph.*, 42(3,4):229–242.

Gelchinsky, B. and Keydar, S. (1999). Homeomorphic imaging approach – theory and practice. *J. Appl. Geoph.*, 42(3,4):169–228.

Gjøystdal, H., Reinhardsen, J. E., and Ursin, B. (1984). Traveltime and wavefront curvature calculations in three-dimensional inhomogeneous media with curved interfaces. *Geophysics*, 49(9):1466–1494.

Gray, S. H. (2001). Seismic imaging. *Geophysics*, 66(1):17–17.

Gray, S. H., Etgen, J., Dellinger, J., and Whitmore, D. (2001). Seismic migration problems and solutions. *Geophysics*, 66(5):1622–1640.

Hertweck, T. (2004). *True-amplitude Kirchhoff migration: analytical and geometrical considerations*. Logos Verlag, Berlin.

Höcht, G. (1998). The Common Reflection Surface Stack. Master's thesis, University of Karlsruhe, Germany.

Höcht, G. (2002). *Traveltime approximation for 2D and 3D media and kinematic wavefield attributes*. PhD thesis, University of Karlsruhe, Germany.

Höcht, G., Bergler, S., Perroud, H., and de Bazelaire, E. (2003). 3D velocity inversion using kinematic wavefield attributes. In *Extended Abstracts*. 65th Mtg., Eur. Assn. Geosci. Eng. Session B-03.

Höcht, G., de Bazelaire, E., Majer, P., and Hubral, P. (1999). Seismics and optics: hyperbolae and curvatures. *J. Appl. Geoph.*, 42(3,4):261–281.

Höcht, G., Perroud, H., and Hubral, P. (1997). Migrating around on hyperbolas and parabolas. *The Leading Edge*, 16(5):473–476.

Hubral, P. (1983). Computing true amplitude reflections in a laterally inhomogeneous earth. *Geophysics*, 48(8):1051–1062.

Hubral, P. (1984). Simulating true amplitude reflections by stacking shot records. *Geophysics*, 49(3):303–306.

Hubral, P. and Krey, T. (1980). *Interval velocities from seismic reflection traveltime measurements*. Soc. Expl. Geophys.

Hubral, P., Schleicher, J., and Tygel, M. (1992a). Three-dimensional paraxial ray properties part I: basic relations. *J. Seis. Expl.*, 1:265–279.

Hubral, P., Schleicher, J., and Tygel, M. (1992b). Three-dimensional paraxial ray properties part II: applications. *J. Seis. Expl.*, 1:347–362.

Hubral, P., Schleicher, J., Tygel, M., and Hanitzsch, C. (1993). Determination of Fresnel zones from traveltime measurements. *Geophysics*, 58(8):703–712.

Jäger, R. (1999). The Common Reflection Surface Stack - Introduction and Application. Master's thesis, University of Karlsruhe, Germany.

Jäger, R., Mann, J., Höcht, G., and Hubral, P. (2001). Common-reflection-surface stack: Image and attributes. *Geophysics*, 66(1):97–109.

Jones, I. F., Ibbotson, K., Grimshaw, M., and Plasterie, P. (1998). 3-D prestack depth migration and velocity model building. *The Leading Edge*, 17(7):897–906.

Kirkpatrick, S., Gelatt, C., and Vehhi, M. (1983). Optimization by simulated annealing. *Science*, 220:671–680.

Landa, E., Gurevich, B., Keydar, S., and Trachtman, P. (1999). Application of multifocusing method for subsurface imaging. *J. Appl. Geoph.*, 42(3,4):283–300.

Levin, F. K. (1971). Apparent velocity from dipping interface reflections. *Geophysics*, 36(3):510–516.

Liu, Z. and Bleistein, N. (1995). Migration velocity analysis: Theory and an iterative algorithm. *Geophysics*, 60(1):142–153.

Loewenthal, D., Roberson, D., and Sherwood, J. (1976). The wave equation applied to migration. *Geophys. Prosp.*, 24(1):380–399.

Mann, J. (2002). *Extensions and applications of the Common-Reflection-Surface Stack Method.* Logos Verlag, Berlin.

Mann, J., Duveneck, E., Hertweck, T., and Jäger, C. (2003). A seismic reflection imaging workflow based on the Common-Reflection-Surface stack. *J. Seis. Expl.*, 12:283–295.

Mann, J. and Höcht, G. (2003). Pulse stretch effects in the context of data-driven imaging methods. In *Extended Abstracts*. 65th Mtg., Eur. Assn. Geosci. Eng. Session P007.

Mann, J., Jäger, R., Müller, T., Höcht, G., and Hubral, P. (1999). Seismics and optics: hyperbolae and curvatures. *J. Appl. Geoph.*, 42(3,4):301–318.

Müller, N.-A. (2003). The 3D Common-Reflection-Surface Stack - Theory and Application. Master's thesis, University of Karlsruhe.

Müller, T. (1999). *The common reflection surface stack - seismic imaging without explicit knowledge of the velocity model.* Der Andere Verlag, Bad Iburg.

Neidell, N. S. and Taner, M. T. (1971). Semblance and other coherency measures for multichannel data. *Geophysics*, 36(3):482–497.

Perroud, H., Hubral, P., de Bazelaire, E., and Höcht, G. (1997). Migrating around in circles – Part III. *The Leading Edge*, 16(6):875–883.

Perroud, H. and Tygel, M. (2004). Nonstretch NMO. *Geophysics*, 69(2):599–607.

Ratcliff, I. F., Ibbotson, K., Grimshaw, M., and Plasterie, P. (1994). Subsalt imaging via target-oriented 3-D prestack depth migration. *The Leading Edge*, 13(3):163–170.

Rietveld, W. and Summers, Y. (2002). Prestack depth imaging: from exploration to production. In *Expanded Abstracts*, pages 1300–1303. 72nd Annual Internat. Mtg., Soc. Expl. Geophys.

Schleicher, J. (1993). *Bestimmung von Reflexionskoeffizienten aus Reflexionsseismogrammen.* PhD thesis, University of Karlsruhe, Germany.

Schleicher, J., Hubral, P., Tygel, M., and Jaya, M. S. (1997). Minimum apertures and Fresnel zones in migration and demigration. *Geophysics*, 62(1):183–194.

Schleicher, J., Tygel, M., and Hubral, P. (1993). Parabolic and hyperbolic paraxial two-point traveltimes in 3D media. *Geophys. Prosp.*, 41(4):495–514.

Schleicher, J., Tygel, M., and Hubral, P. (2004). *Seismic True Amplitude Reflection Imaging.* To be published as SEG Monograph.

Schleicher, J., Tygel, M., Ursin, B., and Bleistein, N. (2001). The Kirchhoff-Helmholtz integral for anisotropic elastic media. *Wave Motion*, 34:353–364.

Schneider, W. (1978). Integral formulation for migration in two and three dimensions. *Geophysics*, 43(1):49–76.

Sheriff, R. E. and Geldart, L. P. (1995). *Exploration Seismology*. Cambridge University Press, Cambridge.

Stork, C. and Clayton, R. (1991). Linear aspects of tomographic velocity analysis. *Geophysics*, 56(4):483–495.

Thore, P., de Bazelaire, E., and Ray, M. P. (1994). The three-parameter equation: An efficient tool to enhance the stack. *Geophysics*, 59(2):297–308.

Trappe, H., Gierse, G., and Pruessmann, J. (2001). Case studies show potential of Common Reflection Surface stack – structural resolution in the time domain beyond the conventional NMO/DMO stack. *First Break*, 19:625–633.

Tygel, M., Müller, T., Hubral, P., and Schleicher, J. (1997). Eigenwave based multiparameter traveltime expansions. In *Expanded Abstracts*, pages 1770–1773. 67th Annual Internat. Mtg., Soc. Expl. Geophys.

Tygel, M., Schleicher, J., and Hubral, P. (1992). Geometrical spreading corrections in a laterally inhomogeneous earth. *Geophysics*, 57(8):1054–1063.

Ursin, B. (1982). Quadratic wavefront and traveltime approximations in inhomogeneous layered media with curved interfaces. *Geophysics*, 47(7):1012–1021.

Vanelle, C. and Gajewski, D. (2002a). Second-order interpolation and traveltimes. *Geophys. Prosp.*, 50(1):73–83.

Vanelle, C. and Gajewski, D. (2002b). True-amplitude migration weights from traveltimes. *Pure Appl. Geophys.*, 159:1583–1599.

Vermeer, G. J. O. (1998). 3-D symmetric sampling. *Geophysics*, 63(5):1629–1647.

Vieth, K.-U. (2001). *Kinematic wavefield attributes in seismic imaging*. PhD thesis, University of Karlsruhe.

Yilmaz, O. (2001a). *Seismic data analysis*, volume 1. Soc. Expl. Geophys., Tulsa.

Yilmaz, O. (2001b). *Seismic data analysis*, volume 2. Soc. Expl. Geophys., Tulsa.

Zhang, Y., Bergler, S., and Hubral, P. (2001). Common-Reflection-Surface (CRS) stack for common-offset. *Geophys. Prosp.*, 49(6):709–718.

Danksagung

Mein besonderer Dank gilt **Prof. Dr. Peter Hubral**. Er hat mich in jeglicher Hinsicht unterstützt und war für mich ein richtiger Doktorvater. Von der Möglichkeit, innerhalb des WIT-Konsortiums Kontakte zu anderen Wissenschaftlern zu knüpfen und meine Arbeit auf internationalen Tagungen präsentieren zu können, werde ich auch über meine Dokorandenzeit hinaus profitieren.

Prof. Dr. Dirk Gajewski danke ich für die Übernahme des Korreferats und die spontane Hilfe bei meinen Terminschwierigkeiten.

Dr. Jürgen Mann bin ich für die jahrelange Zusammenarbeit und die gemeinsame Zeit im Büro und auf Tagungen zu großem Dank verpflichtet. Er hat mich immer wieder mit seinem enormen Wissen über alle möglichen Dinge verblüfft. Darüber hinaus hat er durch viele Diskussionen und das Korrekturlesen entscheidend zum Gelingen der Arbeit beigetragen.

Dr. German Höcht danke ich für die seine Freundschaft, hartnäckige Diskussionen sowie die gemeinsame Zeit in Karlsruhe und Pau. Gerade am Anfang meiner Doktorandenzeit hat mir sein Wissen sehr weitergeholfen.

Dr. Thomas Hertweck hatte immer für alle Fragen ein offenes Ohr, vor allem bei Rechnerproblemen. Von seinem immensen Einsatz für die Arbeitsgruppe konnte auch ich sehr profitieren. Dafür, aber auch für das Korrekturlesen der Arbeit, sei ihm sehr gedankt. Des Weiteren haben mir unsere gemeinsamen Stunden im Hörsaal sehr viel Spaß gemacht.

Alex Müller danke ich für die gute Zusammenarbeit im Rahmen des 3D CRS stacks, Korrekturlesen der Arbeit und unsere nachmittäglichen Kaffeepausen.

Danke an **Eric Duveneck** für zahllose Diskussionen über die Geophysik, die zur Erstellung der Arbeit sehr hilfreich waren. Außerdem möchte ich ihm für die Zeit zusammen im Büro und auf den Tagungen danken.

Christoph Jäger danke ich für die gute Zusammenarbeit im Rahmen der Arbeitsgruppe und das Korrekturlesen der Arbeit.

Many thanks to **Prof. Dr. Martin Tygel** for giving me the opportunity to visit his working group at the University of Campinas, Brazil. The stay was a great pleasure and a rewarding experience. I always enjoyed talking with him about geophysics and other topics. I acknowledge the financial support I received for the stay.

I am grateful to **Prof. Dr. Hervé Perroud** for his generous offer for accommodation in Campinas. I thank him also for many good discussions.

I would like to thank **Prof. Dr. Maria Amélia Novais Schleicher**, **Prof. Dr. Jörg Schleicher**, and **Prof. Dr. Lúcio Tunes dos Santos** for making my time in Campinas so pleasant. Moreover, I wish to thank **Prof. Dr. Jörg Schleicher** for instructive suggestions that helped to improve this thesis.

I thank **Lucas Batista Freitas** for his true friendship and all his support in Brazil. I miss our daily pingado-meetings. Thanks also to **Farid Majana Fang** for his help in Brazil.

I am thankful to **Dr. Evgeny Landa** for his invitation to visit his OPERA group in Pau, the financial support, and his kind hospitality. I learned a lot from the instructive discussions with him. Moreover, I appreciate the opportunity of testing my computer code on real data.

Thanks a lot to **Dr. Elive Manfred Menyoli** for giving me a home in Pau. I really enjoyed the discussions, his sense of humour, and his African cooking.

Thank you also to all others in the OPERA group, especially **Dr. Reda Baina**, **Dr. Raoul Beauduin**, and **Pascale Silberberg**. In this regard, many thanks to **Enrico Zamboni** for his help in migrating the data.

I wish to thank *WesternGeco* for the permission to utilise their data for my thesis. In particular I am grateful to **Dr. Jérôme Guilbot** from *TOTAL* for his help to get the permission.

Dr. Henning Trappe, **Dr. Jürgen Pruessmann** und **Dr. Radu Coman** von *TEEC* danke ich für interessante Diskussionen und Anregungen zum Thema CRS auf verschiedenen Tagungen und WIT-Meetings.

Claudia Payne sei für ihr offenes Ohr bei allen organisatorischen und bürokratischen Problemen gedankt. Mit ihrer freundlichen Art trägt sie entscheidend zur guten Atmosphäre der Arbeitsgruppe bei.

Petra Knopf danke ich für ihre Hilfe bei allen computerspezifischen Problemen. Ebenso danke ich allen Kollegen für die schöne Zeit am Geophysikalischen Institut.

Michelle Di Leo danke ich für ihre endlose Unterstützung und das Korrekturlesen der Arbeit. Besonderer Dank geht auch an meine Eltern für die Unterstützung während meines Studiums.

Lebenslauf

Persönliche Daten

Name:	Steffen Bergler
Geburtsdatum:	30. April 1975
Nationalität:	deutsch
Geburtsort:	Isny im Allgäu

Schulausbildung

1981 - 1985	Grundschule Christazhofem
1985 - 1994	Gymnasium Isny im Allgäu
15.06.1994	Allgemeine Hochschulreife

Hochschulausbildung

1995 - 2001	Studium der Geophysik an der Universität Karlsruhe (TH)
21.02.2001	Diplom
seit 2001	Doktorand an der Fakultät für Physik der Universität Karlsruhe (TH)